# Advances in Atom and Single Molecule Machines

More information about this series at http://www.springer.com/series/10425

Philip Moriarty · Sebastién Gauthier
Editors

# Imaging and Manipulation of Adsorbates Using Dynamic Force Microscopy

## Proceedings from the AtMol Conference Series, Nottingham, UK, April 16–17, 2013

Springer

*Editors*
Philip Moriarty
School of Physics and Astronomy
University of Nottingham
Nottingham
UK

Sebastién Gauthier
CMES-CNRS
Toulouse
France

ISSN 2193-9691 ISSN 2193-9705 (electronic)
Advances in Atom and Single Molecule Machines
ISBN 978-3-319-17400-6 ISBN 978-3-319-17401-3 (eBook)
DOI 10.1007/978-3-319-17401-3

Library of Congress Control Number: 2015936381

Springer Cham Heidelberg New York Dordrecht London


Printed on acid-free paper

Springer International Publishing AG Switzerland is part of Springer Science+Business Media
(www.springer.com)

# Foreword

There are many elegant examples of the manipulation of atoms and molecules on conducting and semiconducting surfaces using the scanning tunnelling microscope (STM), and a number of the chapters in this volume cite those studies. Dynamic force microscopy (DFM, also known as non-contact atomic force microscopy), however, opens up the possibility of constructing atomic-scale circuits, devices and machines on an *insulating* substrate. A key advantage of an electronically insulating surface supported by a robust bulk material is that the leakage current between interconnecting surface metallic contacts (i.e. 'nano-pads') will be very small, suppressing crosstalk between parts of the atomic-scale circuit. Similarly, in order to effectively electronically decouple the device from its environment, the electronic band structure of the bulk material must also be characterised by a large band gap.

These criteria impose important limitations on the choice of the supporting material, the surface preparation techniques (growth, cleavage, and thermal re-structuration and/or reconstruction), the atomic-scale surface characterisation techniques used for analysis, and the final packaging of the circuit (using surface passivation, for example). A crude estimate of the surface tunnelling current intensity leads to the requirement of an electronic band gap larger than 6 eV in order to have surface leakage currents below a few picoamperes (at moderate operating voltages and for inter-contact distances of approximately 10 nm).

With these criteria in mind, and in parallel with the exploration of insulating supporting materials and surfaces, a number of important technological and scientific drivers can be listed. These span, for example, the development of single-molecule machines based on theoretical predictions and the generation of atomic-scale logic circuits at surfaces. The production of wafers comprising these types of devices is also important from an industrial point of view but requires exceptionally advanced processing and sensitive characterisation tools: the wafers must have close-to-perfect atomically ordered terraces. Currently, it is a major technological challenge to produce atomically flat terraces larger than 1 μm across without steps, emerging dislocations, or atomically scale defect. This type of atomically precise surface engineering paramount is important in order to scale-up surface circuit complexity.

The possibility of imaging the surface topography of such ideal surfaces with 'subatomic' scale resolution in real space is a prerequisite for the development of the single molecule and atomic logic gates whose development is pursued within the AtMol project. Traditional, optical, and electronic spectroscopies can still provide valuable information but simply do not have the necessary spatial resolution. Moreover, relatively high-energy electrons and/or photons can induce damage to the surface (and, indeed, of the bulk material below) induced by optical transitions or electronic charging effects. A key consequence has been the resurgence of the physicist's interest in forces for imaging and manipulations at the atomic scale as presented in this, Volume 6 of the series.

Atomic force microscopy (AFM) has now been developed in many directions such as the contact, tapping and non-contact modes of operation, with a wide range of variations in terms of cantilever stiffness, oscillation frequency and amplitude. In the context of the atomic- and sub-molecular-scale manipulation necessary for the type of insulator-supported devices described above, NC-AFM seems now the instrument of choice. Pushed by the performance and the precision reached by the STM in the manipulation of single atoms and molecules and in the construction of atomic-scale circuits and molecular machineries, scientists developing NC-AFM instruments have an important benchmark in place to push force microscopy (and its associated spectroscopic capabilities) to the limits. They can provide the characterisation tools and manipulation protocols at the atomic scale that have been lacking for years at the surface of an insulator. Some research groups have also tried to push the STM to the frontier of the field emission regime, succeeding in imaging large electronic gap surfaces and single molecules on a metallic surface using tip-to-surface bias voltages of up to 10 V. This is very much at the limit of what an STM can do on insulating surfaces. Given that it is operating right at the limits of its capabilities, this application of the STM is not very practical for the atomic-scale operations required for structures and devices constructed or assembled at the surface of large bandgap materials.

The demonstrated ability to image an adsorbate on a large bandgap material is a superb scientific avenue not only in terms of acquiring detailed information on the electronic structure of a single atom or molecule but, moreover, to provide detailed access 'inside' the electronic cloud of the adsorbate (i.e. to image the atomic 'architecture' of a molecule). Even if the description of a molecule by its valence electronic density probability is a long-standing and well-illustrated feature of many textbooks, the recording of real-space experimental images of the 'internal structure' of an adsorbate is still a matter of considerable debate. Using a large electronic gap substrate is one way to minimise the direct electronic hybridisation between the atom, the molecule and its supporting surface. Strong hybridisation is very often encountered on metal and semiconductor surfaces. Here, the large number of atoms of the surface involved in strong electronic interactions with the adsorbate in essence act like a 'quantum-to-classical' convertor for the adsorbate. This coupling to the environment removes a key part of the quantum richness of the intrinsic electronic structure of, for example, a molecular adsorbate. In this context, measuring with 'subatomic' resolution the very local force between the end apex atom of an AFM tip and the molecule's electronic cloud provides an entirely different method of accessing the internal electronic structure of the molecule as compared, for example, with field emission microscopy (FEM), transmission electron microscopy (TEM), or STM. The binding of single vertical molecule like CO to the apex of a scanning probe, as shown by Gross et al. at IBM Zurich, produces a fantastic force sensor to probe the interior of the electronic cloud of a molecule. This apex can also, in principle, be equipped with a high-spin atom for magnetic interrogation of the molecule's electronic structure opening up the possibility of intramolecular spin mapping.

The fundamental rationale behind those measurements lies in the problem of representing the very complex electronic structure of a many-electron molecule. Can the representation of such a complex electronic system be simply reduced to a pseudo-3D map of the ground electronic state of the molecule? Physisorbed at the surface of an oxide, a molecule is largely an isolated (but supported) minute piece of matter. At low temperature, its mechanical degrees of freedom will have some (or perhaps even total) classical behaviour due to quantum decoherence. This is a remarkable platform (or playground) for performing single-molecule mechanics, developing molecule motors and constructing molecular machines on a surface where the relation between the insulator characteristics and friction at the atomic scale remains to be explored. What role do the electronic degrees of freedom play in such a supported (or weakly coupled) situation? How can the corresponding many-body wave function be mapped (with local 'subatomic' resolution)? Is this local force spatial mapping closer in information content to the famous FEM images of a single isolated molecule (at the end of a tungsten tip) acquired by Muller in the mid-fifties, or to the direct space molecular orbital like images obtained by $dI/dV$ constant current STM images? How might magnetism and the local mechanical response of a molecule be related? These are just a few questions posed by our new ability to probe the interior of the electronic cloud of a weakly adsorbed/interacting molecule.

A key additional advantage of an insulating surface is that it can be 'equipped' locally by nanoscale objects such as semiconductor nano-clusters, nanoscale metallic contacts ('nano-pads'), and even more functionalised nano-devices such as nano-gears. For example, ultraflat metallic nano-pads can be assembled directly on an insulating surface using a source of metal atoms or by transferring graphene nano-islands on the surface (with always the same constraint that these nano-islands must be defined with atomic-scale precision). Specific UHV growth and transfer printing techniques have now been developed for this purpose which preserve the atomic-scale order of the surface. NC-AFM can also play a key role in reconfiguring the surface location of adsorbed nano-objects.

One interesting phenomenon is the electronic interaction of an adsorbate with a single metallic nano-pad. The challenge here is the study of electron transfer and the consequent charging of a single atom or a single molecule at the atomic scale by a nearby metallic nano-island previously charged by simple electrostatic contact. To what extent, at the surface of an insulator, does this type of phenomenon occur? What is the lifetime of this charging effect and how does it depend on the large electronic band gap of the supporting material? Aside from these electrostatic and electronic effects, mechanical interactions (such as the gearing between a molecule gear and a solid-state nano-gear) will also benefit from the use of an insulating surface to minimise the electronic friction encountered at the surface of, for example, a metal.

Following the investigation of the influence of an individual nanoscale object, the next logical step is the study of the interactions of atomic or molecular adsorbates with a number of nano-clusters. This will open up vast field of research for nano-electronic, nano-mechanics and, very soon, nano-magnetism and nano-optics (nano-plasmonics), which all will reach, in the years to come, atomic-scale precision. The adsorbate can be a single-molecule machine able to calculate, to 'memorise' data, or to perform mechanical work. By extension, it could also be an atomic-scale electronic circuit or complex mechanical machinery. In all these cases, avoiding through-surface inter-contact leakage current will remain a crucial objective.

The applied bias voltage required to drive nanoscale machinery or atomic-scale circuits via metallic nanoscale contacts will necessarily be smaller than the 6 eV band gap mentioned above. Having such a fully planar surface interconnection arrangement is the next frontier for the realisation of atomic-scale machines and technology. This is so crucial that this ongoing planarisation of atomic-scale technology is now triggering the development of a new class of instruments mixing near-field and far-field technologies and microscopies. These new microscopes, combining four low-temperature ultrahigh vacuum STMs or four NC-AFMs (or, indeed, a mix of both instruments) are now able to image and manipulate at the atomic scale in parallel and on the same surface. Such instruments incorporate an SEM, and sometimes a focused ion beam system, to ensure accurate navigation of the apices of four probes with respect to other nanoscale objects. With this new development, it is now the instrumentation which is adapted to the scale of the atomic machinery, rather than the use of 'scale-up' protocols (such as crystallisation) to enable the formation of a material compatible with the standard far-field characterisation and interconnection techniques.

These advances in instrumentation are underpinning the evolution of atomic-scale devices, machines and circuits towards a planar geometry. This planarisation in turn both opens a new window for experimentation and puts new challenges on the table including, in particular, the description of the observed phenomena using modern calculations tools. The combination of a large electronic gap surface, a bulk material as support, solid-state nanoscale structures for contacts and interconnects, and active single-molecule machines or atomic-scale circuit is a redoubtable challenge for quantum chemists. How can all these scales be encompassed, while also taking into account the precise electronic structure of each atom? Can it ever be only a semi-empirical description? What about the time-dependent responses of the atomic/molecular machinery? Can this be predicted from first principles before setting up the required instrumentation?

These exceptionally challenging questions have been largely outside the scope of the STM, for which metal and semiconductor surfaces were exploited for atomic-scale construction and for the development of the first elementary molecule machines or atomic-scale circuits. The introduction of large bandgap surfaces, and the current availability of exceptionally high-precision non-contact AFM instrumentation, represent an exciting opportunity to experimentally realise single-molecule machines and atomic circuits previously only developed 'in silico'.

To finish with, let me thank P. Moriarty and S. Gauthier for their time and patience in organizing the April 2013 AtMol workshop on non-contact atomic force microscopy-related techniques and all the participants of this workshop to have accepted to participate to this Volume 6.

Nottingham
April 2013

C. Joachim

# Preface

This book, like others in the *Advances in Atom and Single Molecule Machines* series, stems from a two-day workshop associated with the Atomic Scale and Single Molecule Logic Gate Technologies (AtMol) project funded by the European Commission's Future and Emerging Technologies programme. AtMol, a four-year project, was launched on 1 January 2011, involves eleven research teams, and has as its central objective the design and fabrication of the first ever encapsulated single-molecule chip.

Scanning probe microscopy is at the very heart of the AtMol project, enabling the imaging and manipulation of matter at the atomic, molecular, and sub-molecular scales. As such, the workshop from which the chapters in this volume arise was funded both by AtMol and the ACRITAS Initial Training Network (www.acritas.eu), a Marie Curie Training Project also funded by the European Commission and which is focussed on the development of new techniques and methods in atomic force microscopy.

The workshop in question, *Imaging and Manipulating of Adsorbates Using Dynamic Force Microscopy*, was held at the University of Nottingham during 16–17 April 2013. The contributions included in this volume represent a 'snapshot' of the state of the art in dynamic force microscopy (DFM, also called non-contact atomic force microscopy) at that time and cover a variety of fascinating topics including the accurate extraction of tip–sample forces from measured frequency shift curves; the imaging and manipulation of single atoms, molecules, clusters, and dangling bonds; the role of the Pauli exclusion principle in high-resolution DFM; and the combined mechanical and electrical properties of individual molecules.

At the time of writing, DFM is at an intriguing juncture in the evolution of the capabilities of the technique. There has been significant excitement about the possibility of imaging intramolecular, and, in particular, inter-molecular bonds, over the last 18 months or so. It has now become clear that the latter, i.e. the observation of contrast maxima *between* molecules in DFM images, has to be interpreted very carefully indeed. As with any scanning probe technique, the geometric and electronic structure of the apex of the tip (and its associated chemical functionality) is just as important in the imaging mechanism as the sample. This convolution can be particularly severe in the so-called Pauli exclusion regime of imaging which was pioneered by IBM Zurich (Gross et al.) in 2009 (and which is increasingly becoming the norm for the acquisition of ultrahigh resolution DFM images).

As one of us (PM) and co-authors discuss in the chapter "Pauli's Principle in Probe Microscopy" of this volume, the contribution of the dynamics and mechanics of the tip apex plays an exceptionally important role in generating image contrast. Seeing is not necessarily believing with any imaging technique, and this is especially true of scanning probe microscopy. On the other hand, there remains immense potential for the use of a range of different functional probes (including, but also moving beyond, CO, which is used extensively for intra molecular imaging at present) in DFM, enabling analysis of a variety of physicochemical phenomena. These include molecular magnetism, as also mentioned in the Foreword to this book. The exciting challenges and opportunities for ultrahigh resolution DFM will be discussed and debated in a workshop in Prague in February 2015 which has been organised by Pavel Jelinek, author of chapter "Theoretical Challenges of Simultaneous Nc-AFM/STM Experiments" of this volume (see below).

In the chapter "Mechanical and Electrical Properties of Single Molecules", Thilo Glatzel provides a comprehensive overview of a number of the considerable advances of the Nanolino group at the University of Basel in imaging and characterising the mechanical and electrical properties of single molecules. For all of the reasons outlined by Christian Joachim in his Foreword, there are fundamental and technologically important motivations for the use of insulating, as opposed to metal or semiconductor, substrates as a 'platform' for molecular adsorption. Glatzel's overview gives a good indication of the state of the art in the field and also highlights key differences between the properties of molecular adsorbates on metallic and insulating substrates.

Sugimoto and colleagues at Osaka University have been responsible for many of the key advances in atomic manipulation and single atom control using DFM. Prof. Sugimoto's chapter reviews some of the work of the Osaka group (and their collaborators) in developing a number of fascinating and high-precision protocols for the controlled positioning of single atoms, with a particular focus on the lateral translation of silicon adatoms. Sugimoto discusses how the tip apex in a DFM manipulation experiment can be characterised through measurement of the short-range chemical force and, in particular, by using the maximum attractive force as a 'metric' or a diagnostic. In addition, however, he stresses how tip geometry—asymmetry in particular—plays a central role in determining the probability for manipulation events to occur.

In a second contribution from the Nottingham group,[1] Andrew Stannard and Adam Sweetman explore the difficulties and subtleties associated with extracting accurate quantitative force measurements from dynamic force microscopy measurements of frequency shift versus tip–sample separation. Although DFM is an

[1]Disclaimer: one of the editors (PM) is a member of this group.

exceptionally powerful tool for the measurement of tip–sample force fields and potential energy landscapes, it is far from trivial to extract accurate and reliable quantitative force values from the primary experimental observable, i.e. the frequency shift ($\Delta f$). Subtraction of the contribution of the long-range van der Waals interaction, in which the short-range chemical force curve is 'buried', is a perennial, and particularly thorny, issue. Stannard and Sweetman describe protocols for the systematic treatment of $\Delta f(z)$ curves (where $z$ represents the tip–sample separation) so as to minimise the uncertainties in converting the experimental data to a plot of $F_{SR}(z)$, i.e. the short-range force curve associated with the interaction of just the atom/molecule terminating the tip apex with the underlying sample surface atom(s).

The chapter "Theoretical Challenges of Simultaneous Nc-AFM/STM Experiments", authored by Pavel Jelinek of the Academy of Sciences of the Czech Republic, is an authoritative and fascinating review of the current state of the art in theoretical descriptions of 'hybrid' scanning tunnelling microscopy (STM)—DFM measurements. Jelinek's group has carried out pioneering and influential work on this topic, and he provides an accessible and readable introduction—perhaps written with experimentalists in mind—of the latest developments in the use of density functional theory, coupled with a Green's function[2] approach, to simulate not only the geometric and chemical interactions occurring at the tip–sample junction, but to accurately predict the (far from equilibrium) tunnel current driven through the system by the application of a bias voltage. The question of just how the distance dependence of the tunnel current compares to that of the short-range chemical force has been a long-standing issue in scanning probe microscopy, and Jelinek discusses the key theoretical advances that have elucidated the underlying physics. He also highlights the fascinating, and 'far beyond perturbative', influence of the formation of a covalent bond on the tunnelling barrier experienced by electrons flowing from tip to sample (or vice versa).

The next chapter, written by Clemens Barth (CNRS, Aix-Marseille University), explores the use of DFM to manipulate objects rather larger than the single atom/molecule adsorbates discussed thus far: nanoscale clusters. As discussed in the Foreword, nanoscale clusters and nanoparticles have a key role to play in enabling the development of next-generation atomic circuitry, particularly with regard to interfacing single atom and sub-molecular functionality to the microscopic, and ultimately macroscopic, world. Barth describes the challenges underpinning controlled translation of clusters and discusses how his group has developed techniques to control the positioning and dynamics of Au nanoparticles on NaCl substrates. The importance of the wide electronic band gap of insulating surfaces and substrates was also stressed repeatedly by Joachim in the Foreword, and it is encouraging that Barth shows that not only is nanoparticle manipulation on wide bandgap substrates achievable, but that the presence of surface defects does not

[2]Given the venue for the workshop from which this volume stemmed, it would be remiss of me not to highlight that George Green, of Green's function fame, did his mathematics while working as a miller in Nottingham in the nineteenth century.

preclude controlled positioning. Indeed, the chapter "Manipulation of Metal Nanoparticles on Insulating Surfaces" provides compelling evidence that defects are an integral part of the manipulation process, first stabilizing, and then acting in concert with the nanocluster as it is moved across the surface by the DFM tip.

In the chapter "Imaging of Defects on Ge(001):H by Non-contact Atomic Force Microscopy", we remain with the theme of defect-mediated interactions but return to the imaging and quantitative analysis of semiconductor, rather than insulator, surfaces. Bartosz Such (Jagiellonian University, Kraków) focuses on a prototypical passivated, i.e. low free energy, semiconductor surface—hydrogen-terminated Ge (100)—and describes the significant differences between force curves acquired over passivated surface regions as compared to those measured above single dangling bond defects (i.e. regions of the surface where a Ge atom is not hydrogen 'capped'). As has also been observed for the H:Si(100) surface, the chemically inert hydrogen-passivated regions are associated with a negligible attractive interaction with the DFM tip. As such, atomic resolution imaging is achieved within the range of the tip–sample interaction potential where the repulsive component of the total force plays a significant role (as discussed in the chapter "Pauli's Principle in Probe Microscopy" in relation to the Pauli exclusion principle). Above a dangling bond defect, however, there is a much stronger attractive force due to the short-range tip–sample interaction arising from the formation of a chemical (presumably covalent) bond. Such discusses how the force of this single dangling bond can be measured.

We subsequently return to a discussion of insulating surfaces in the chapter "Adsorption Structures of Amino Acids on Calcite(104)" but here the focus is on calcite rather than the alkali halides exploited by Barth et al. for the nano-cluster manipulation described in the chapter "Manipulation of Metal Nanoparticles on Insulating Surfaces". As Felix Kling, Markus, Kittelmann and Angelika Kühnle discuss, calcite is a fascinating and exceptionally important substrate in the context of the study of biomineralisation. They exploit the cleavage plane of calcite, i.e. calcite(104), as a platform and template for the self-assembly of five different amino acids. Although the larger molecules of those they studied, namely tryptophan, tyrosine, and aspartic acid, exhibit strikingly similar behaviour on the calcite surface with regard to the superstructures they adopt, the smaller amino acids (glycine and alanine) behave in a very different manner. A rather elegant aspect of Kling and co-worker's study is that they unambiguously define molecular chirality from a careful analysis of the relationship of the alignment of adsorbed superstructure with the crystallographic directions of the underlying substrate.

The final chapter of this volume, by Antoine Hinaut and colleagues, continues with the topic of molecular adsorption on an insulating surface but has a specific focus on the electrostatic properties of the adsorbed molecules (which, like the amino acids studied by Kling et al., form highly ordered superstructures on the substrate). Hinaut et al. have carried out a combined DFM and Kelvin probe force microscopy (KPFM) study of a triphenylene derivative, 2,3,6,7,10,11-hexacyano-propyl-oxytriphenylene, on a KBr surface. A systematic, comparative, and detailed study of the KPFM data acquired above the bare KBr surface and on the molecular islands showed that the Kelvin probe signal, which provides a measure of the local work function of the sample, could be interpreted in terms of the conformation-dependent polarisation of the molecular overlayer. Classical electrostatic calculations of the KPFM signal, based on a spherical tip apex, were found to provide good agreement with the experimental data. Given that there has been considerable discussion and debate regarding the origin and physicochemical underpinnings of KPFM measurements, Hinaut et al.'s work provides key insights into the extent to which intuition and models from classical physics can be used to interpret Kelvin probe data.

Taken together, the nine chapters in this volume capture the state of the art in dynamic force microscopy at an intriguing and exciting time in the evolution of the field. Five years on from the pioneering work of Leo Gross and colleagues at IBM Zurich, who showed that the 'architecture' of adsorbed molecules could be resolved in exquisite detail, there remain a variety of exciting questions to be addressed about the ultimate limits of DFM. Christian Joachim has touched on some of these open questions in his Foreword, and we echo his comments regarding the possibility of measuring local spin density and pushing the magnetic measurement capabilities of the instrument, as explored by Schwarz, Wiesendanger and colleagues at Hamburg, to wider, more challenging, horizons. The recent availability of commercial instruments capable of carrying out DFM in relatively high magnetic fields will of course provide an impetus for this direction of research.

Developments in instrumentation will clearly proceed apace—Franz Giessibl, whose farsighted introduction of the qPlus sensor (followed by the commericalisation of this technology by companies such as Omicron Nanotechnology (now part of Oxford Instruments) and Createc) has driven the wide adoption of DFM in labs across the world, continues to innovate with regard to both sensor design and the application of the qPlus technique to exciting scientific problems. As we write this in early 2015, DFM remains a very slow technique—the effective image generation bandwidth is sub-Hz rather than kHz or MHz. There thus remains particularly exciting scope for the development of DFM instruments capable of providing much greater temporal resolution, to complement the impressively high level of spatial resolution that is now possible with DFM.

We thank the European Commission's ICT-FET and Marie Curie programmes for their financial support in organizing the joint ATMol–ACRITAS workshop and the staff of Springer Verlag (particularly Judith Hinterberg) for their help in producing this book. We are also very grateful to Christian Joachim for leading the AtMol project and for his advice and patience throughout the preparation of this book. Finally, we would like to thank all of the authors who have contributed their work to this volume.

Nottingham<br>Toulouse<br>January 2015

Philip Moriarty<br>Sebastién Gauthier

# Contents

# Contributors

**M. Abe** Graduate School of Engineering, Osaka University, Toyonaka, Osaka, Japan

**Clemens Barth** CNRS, Aix-Marseille Université, Marseille Cedex 09, France

**Janusz Budzioch** Faculty of Physics, Astronomy and Applied Computer Science, Centre for Nanometer-Scale Science and Advanced Materials (NANOSAM), Jagiellonian University, Kraków, Poland

**Florian Chaumeton** CNRS, Centre d'Elaboration des Matériaux et d'Etudes Structurales (CEMES), Toulouse, France

**Sebastién Gauthier** CNRS, Centre d'Elaboration des Matériaux et d'Etudes Structurales (CEMES), Toulouse, France

**Thilo Glatzel** Department of Physics, University of Basel, Basel, Switzerland

**Szymon Godlewski** Faculty of Physics, Astronomy and Applied Computer Science, Centre for Nanometer-Scale Science and Advanced Materials (NANOSAM), Jagiellonian University, Kraków, Poland

**Antoine Hinaut** CNRS, Centre d'Elaboration des Matériaux et d'Etudes Structurales (CEMES), Toulouse, France

**S.P. Jarvis** School of Physics and Astronomy, University of Nottingham, Nottingham, UK

**P. Jelínek** Institute of Physics, Academy of Sciences of the Czech Republic, Prague, Czech Republic

**L. Kantorovich** Department of Physics, Kings College London, London, UK

**Markus Kittelmann** Department of Chemistry, Institute of Physical Chemistry, Johannes Gutenberg-University Mainz, Mainz, Germany

**Felix Kling** Department of Chemistry, Institute of Physical Chemistry, Johannes Gutenberg-University Mainz, Mainz, Germany

**Marek Kolmer** Faculty of Physics, Astronomy and Applied Computer Science, Centre for Nanometer-Scale Science and Advanced Materials (NANOSAM), Jagiellonian University, Kraków, Poland

**Angelika Kühnle** Institute of Physical Chemistry, Department of Chemistry, Johannes Gutenberg-University Mainz, Mainz, Germany

**Jakub Lis** Faculty of Physics, Astronomy and Applied Computer Science, Centre for Nanometer-Scale Science and Advanced Materials (NANOSAM), Jagiellonian University, Kraków, Poland

**David Martrou** CNRS, Centre d'Elaboration des Matériaux et d'Etudes Structurales (CEMES), Toulouse, France

**E. McGlynn** School of Physical Sciences, Dublin City University, Glasnevin, Dublin 9, Ireland

**P. Moriarty** School of Physics and Astronomy, University of Nottingham, Nottingham, UK

**S. Morita** The Institute of Scientific and Industrial Research, Osaka University, Ibaraki, Osaka, Japan

**Andrew Stannard** School of Physics and Astronomy, University of Nottingham, Nottingham, UK

**Bartosz Such** Faculty of Physics, Astronomy and Applied Computer Science, Centre for Nanometer-Scale Science and Advanced Materials (NANOSAM), Jagiellonian University, Kraków, Poland

**Y. Sugimoto** Graduate School of Engineering, Osaka University, Suita, Osaka, Japan

**Adam M. Sweetman** School of Physics and Astronomy, University of Nottingham, Nottingham, UK

**Marek Szymonski** Faculty of Physics, Astronomy and Applied Computer Science, Centre for Nanometer-Scale Science and Advanced Materials (NANOSAM), Jagiellonian University, Kraków, Poland

**Mateusz Wojtaszek** Faculty of Physics, Astronomy and Applied Computer Science, Centre for Nanometer-Scale Science and Advanced Materials (NANOSAM), Jagiellonian University, Kraków, Poland

# Pauli's Principle in Probe Microscopy

**S.P. Jarvis, A.M. Sweetman, L. Kantorovich, E. McGlynn and P. Moriarty**

> *It appears to be one of the few places in physics where there is a rule which can be stated very simply, but for which no one has found a simple and easy explanation. The explanation is deep down in relativistic quantum mechanics. This probably means that we do not have a complete understanding of the fundamental principle involved.*
>
> RP Feynman, *The Feynman Lectures on Physics*, Vol III, Chapter 4 (1964).

**Abstract** Exceptionally clear images of intramolecular structure can be attained in dynamic force microscopy (DFM) through the combination of a passivated tip apex and operation in what has become known as the "Pauli exclusion regime" of the tip–sample interaction. We discuss, from an experimentalist's perspective, a number of aspects of the exclusion principle which underpin this ability to achieve submolecular resolution. Our particular focus is on the origins, history and interpretation of Pauli's principle in the context of interatomic and intermolecular interactions.

**Keywords** Dynamic force microscopy · Non-contact atomic force microscopy · NC-AFM · Pauli exclusion principle · Submolecular resolution · Intramolecular · Single-molecule imaging

S.P. Jarvis · A.M. Sweetman · P. Moriarty (✉)
School of Physics and Astronomy, University of Nottingham, Nottingham NG7 2RD, UK
e-mail: philip.moriatry@nottingham.ac.uk

L. Kantorovich
Department of Physics, Kings College London, The Strand, London WC2R 2LS, UK

E. McGlynn
School of Physical Sciences, Dublin City University, Glasnevin, Dublin 9, Ireland

 

P. Moriarty and S. Gauthier (eds.), *Imaging and Manipulation of Adsorbates Using Dynamic Force Microscopy*, Advances in Atom and Single Molecule Machines,
DOI 10.1007/978-3-319-17401-3_1

# 1 Intramolecular Resolution via Pauli Exclusion

In 2009, the results of a pioneering dynamic force microscopy (DFM[1]) experiment by Leo Gross and co-workers at IBM Zürich were published [1] and revolutionised the field of scanning probe microscopy. Gross et al. captured arguably the clearest real space images of a molecule achieved up to that point, resolving the "textbook" structure of the molecular architecture. Two important experimental protocols enabled Gross et al.—and, subsequently, a number of other groups [2–8] (see Fig. 1 for examples)—to attain this exceptionally high resolution. First, the apex of the probe was functionalised (by picking up a molecule) to render it inert. This enabled the scanning probe to be placed extremely close to the adsorbed molecule of interest—so close that the second experimental protocol, namely the exploitation of electron repulsion via the Pauli exclusion principle (PEP),[2] played a key role in the imaging mechanism.

It is this second protocol which is the primary focus of this chapter. We will discuss just how Pauli exclusion is exploited in state-of-the-art scanning probe microscopy, what pitfalls there might be in interpreting features in DFM images as arising directly from chemical bonds, and to what extent scanning probe measurements of tip–sample interactions provide deeper experimental insights into the exclusion principle itself. We should also stress right from the outset that although we concentrate on DFM throughout this chapter, prior to Gross et al.'s 2009 paper, Temirov, Tautz and co-workers had achieved unprecedented spatial resolution using a technique for which they coined the term scanning tunnelling hydrogen microscopy (STHM) [9–12]. Both STHM and thetype of DFM imaging introduced by Gross et al. [1] exploit Pauli exclusion as a means to acquire exceptionally high resolution. Before covering the exploitation of the exclusion principle in scanning probe microscopy, we will consider a number of aspects of the fascinating history of Pauli's *Ausschließungsregel* [13] and outline some of the rich physics underpinning the principle.

# 2 A Potted History of the Pauli Exclusion Principle

Michela Massimi has written an authoritative and engaging history of the PEP [13], which impressively combines clear explanations of the quantum and statistical physics underlying the PEP with engaging discussions of both the history and the

[1]Although the term non-contact atomic force microscopy (NC-AFM) is widespread—to the extent that the major conference in the field is the annual International NC-AFM meeting—it is arguably something of a misnomer to label the technique "non-contact" when it is now commonplace to operate in a regime where the probe is in contact with the sample. We will therefore use the term *dynamic* force microscopy throughout this chapter.

[2]We shall return, in Sects. 4 and 5, to a detailed discussion of whether or not it is appropriate to describe the effects of Pauli exclusion as a repulsive force.

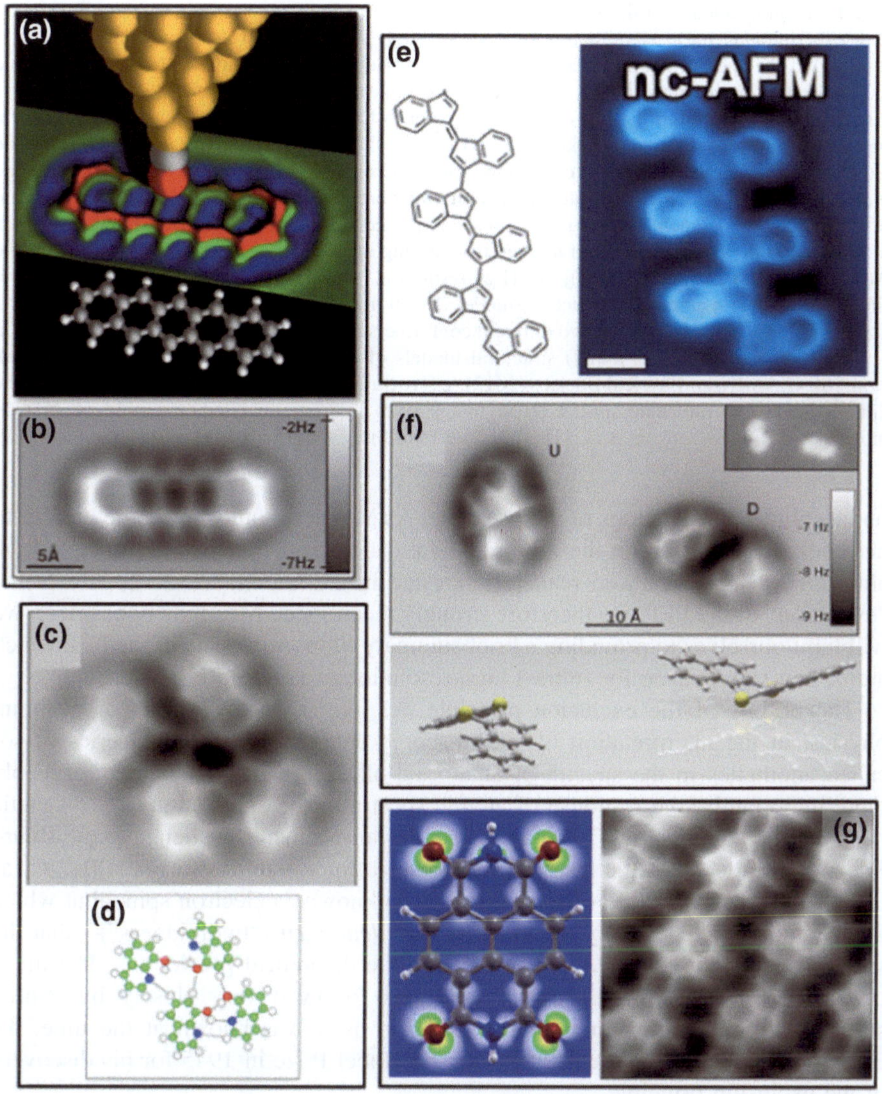
(a)
(b)
-2Hz
-7Hz
5Å
(c)
(d)
(e)
nc-AFM
(f)
U
D
-7 Hz
-8 Hz
-9 Hz
10 Å
(g)

◀ **Fig. 1** Imaging bonds via the Pauli exclusion principle. a Combination of schematic illustration and experimental data to demonstrate experimental protocol used to acquire submolecular resolution. The apex of the probe of a dynamic force microscope is passivated (in this case with a CO molecule) and scanned across a pentacene molecule at a height where Pauli exclusion plays a key role in determining the tip–sample interaction. **b** Experimental frequency shift image for a pentacene molecule [**a** and **b** taken from Gross et al. [1]. © American Association for the Advancement of Science (2009)]. **c** Dynamic force microscope image of four 8-hydroxyquinoline molecules. Both intra- and intermolecular features are observed (See Sect. 7). **d** Schematic diagram of molecular arrangement shown in **c** with the expected positions of hydrogen bonds drawn as lines between the molecules [**c** and **d** taken from Zhang et al. [3]. © American Association for the Advancement of Science (2012)]. **e** High-resolution image of a chain of oligo-(E)-1,1-bi (indenylidene) with associated structural model. Taken from Riss et al. [8]. © American Chemical Society (2014). **f** DFM image of two different conformers of dibenzo[a,h]thianthrene on a NaCl/Cu (111) substrate with (*lower panel*) structural models of both conformers. Taken from Pavlicek et al. [6]. © American Physical Society (2012). **g** Structural model of a naphthalenetetracarboxylic diimide (NTCDI) molecule and a DFM image of a hydrogen-bonded assembly of NTCDI molecules. From Sweetman et al. [4]. © Nature Publishing Group (2014)

philosophical ramifications of the principle. As Massimi points out in the preface to her book, her research on the origin and validation of the exclusion principle took almost 10 years. For those readers interested in a comprehensive account of the "evolution" of the PEP, we therefore strongly recommend Masimi's book. Here, we will limit ourselves to providing a brief summary of those aspects of the PEP which are of key significance for (ultra-) high-resolution scanning probe microscopy.

The origins of the exclusion principle lie, like so many aspects of quantum physics, in the interpretation of spectroscopic data. In particular, a series of so-called anomalies in the spectra of alkali and alkaline earth metals, and, arguably more importantly, the response of atomic spectra to the application of a magnetic field, i.e. the ("anomalous") Zeeman effect, became a major challenge to the Bohr–Sommerfeld theory of the electronic structure of atoms in the early 1920s. It was only with the introduction of what came to be known as electron spin—but which Pauli initially called simply the electron *Zweideutigkeit* ("twofoldness")—that the spectroscopic data could be reconciled with the theoretical predictions. The introduction of electron *Zweideutigkeit* [14] was followed very closely by Pauli's statement of the exclusion principle [15] (or, as it was known at the time, the exclusion rule). Pauli subsequently won the Nobel Prize in 1945 for his discovery of the exclusion principle.

It is worth quoting directly from Pauli's Nobel Lecture, given on 13 December 1946, as this provides key insights into the original formulation of the principle "straight from the horse's mouth", as it were:

> On the basis of my earlier results on the classification of spectral terms in a strong magnetic field the general formulation of the exclusion principle became clear to me. The fundamental idea can be stated in the following way:
>
> The complicated numbers of electrons in closed subgroups are reduced to the simple number *one* if the division of the groups by giving the values of the four quantum numbers of an electron is carried so far that every degeneracy is removed. An entirely non-degenerate

> energy level is already *closed*, if it is occupied by a single electron; states in contradiction with this postulate have to be excluded.

Or, if we couch this in the lexicon of modern quantum mechanics, no two electrons can have the same values of $n$, $l$, $m_l$ and $m_s$ (i.e. the principal, orbital angular momentum, magnetic and spin quantum numbers). More succinctly, no two electrons can occupy the same quantum state. (The PEP of course holds for all fermions (half-integer spin particles), not just electrons. We will return to this point very soon).

Pauli's *Zweideutigkeit* is now of course known as particle spin, but the inferred connection with the classical concept of a spinning object is unfortunately misleading. Indeed, Pauli himself switched from being firmly opposed to any connection between his *Zweideutigkeit* and spin, to a somewhat grudging acceptance of a link, and then, as his Nobel lecture highlights, back to a significant degree of scepticism about the value of any classical analogy:

> On the other hand, my earlier doubts as well as the cautious expression "classically non-describable two-valuedness" experienced a certain verification during later developments, since Bohr was able to show on the basis of wave mechanics that the electron spin … must therefore be considered as an essentially quantum mechanical property of the electron.

## *2.1 Particle Statistics and the Quantum Identity Crisis*

Following hot on the heels of Pauli's publication of the exclusion principle, first Fermi [1, 16, 17] and then Dirac [18] explored the quantum statistics of an ideal gas of particles which was subject to the exclusion principle. Dirac coined the term *fermion* to describe a particle subject to the Fermi–Dirac statistics he and Fermi derived; a fermion is therefore a particle which obeys the PEP (and concomitantly is of half-integer spin). At the very heart of quantum statistics—and, indeed, of classical statistical mechanics—lies the issue of the distinguishability of particles.[3] A simple back-of-the-envelope argument based on the (in) distinguishability of particles can provide a helpful insight into the origin of the exclusion principle [19].

Before we introduce that back-of-the-envelope approach, however, it is first important to define just what it is we mean by indistinguishable particles. This, despite first appearances, is a far from trivial question to address and has been the subject of quite considerable debate and interest for many decades. De Muynck [20], Berry and Robbins [21], Ginsberg et al. [22] (see also Fleischhauer [23] for a very readable overview of Ginsberg et al.'s work), Omar [24], and Dieks and co-workers [1, 25, 26], amongst many others, have considered and explored the important issue of how indistinguishability and quantum statistics are intrinsically

[3] Long before the advent of quantum mechanics, the effect of considering indistinguishable versus distinguishable particles on the partition function for a system was known as the Gibbs paradox in classical thermodynamics/statistical mechanics.

coupled. We shall not delve into the detailed arguments—be they physical, philosophical or semantic in scope—and instead restrict ourselves to the following relatively simple, although certainly not "universal", definitions. (It is also important to note that the condition for antisymmetry and the exclusion principle are not equivalent statements).

First, we draw a distinction between *identical* and *indistinguishable* particles. Identical particles are those which have the same intrinsic (or "internal") properties (and the same values associated with those intrinsic properties), i.e. mass, charge and spin. So two electrons are identical to each other. And two protons, or two neutrons, are similarly identical to each other. But electrons are clearly not identical to protons, nor to neutrons. (We apologise for labouring the point to this extent, but the terms "identical" and "indistinguishable" are often used interchangeably—including in many textbooks—and this has led to quite some confusion at times).

If we have a collection of identical particles, then they are *indistinguishable* if we cannot separate them on the basis of their "external" properties such as position or momentum. But classically, it is possible to distinguish between identical particles (at least in principle): we can effectively "label" individual identical particles on the basis of their positions or trajectories and distinguish them accordingly.[4] Quantum mechanically, however, the standard argument is that due to delocalisation, we lose this ability to label particles on the basis of their trajectories and they then become indistinguishable.

But to what extent is this true? Are quantum particles indeed indistinguishable? One can find undergraduate-level descriptions of quantum statistics [30] which claim that quantum particles can in fact be distinguished on the basis of what might be called a "Rayleigh criterion" for wave packets: if two particles are separated by a distance greater than their de Broglie wavelength (i.e. such that the wave function overlap is minimal), then they are distinguishable on the basis of their respective positions. Versteegh and Dieks [27] invoke similar arguments about the spatial extent of wave packets enabling identical quantum particles to be distinguished.

However, whether this is a valid condition for distinguishability is far from clear-cut. In his commentary on Ginsberg et al.'s work [22], Fleischhauer [23] states the following:

> In the quantum world, particles of the same kind are indistinguishable: the wave function that describes them is a superposition of every single particle of that kind occupying every

[4]In a thought-provoking paper, Versteegh and Dieks [27] discuss the importance of the distinguishability of identical particles and what these mean for classical thermodynamics and statistical mechanics, including the Gibbs paradox. We note, however, that there is a very important omission in the list of papers cited by Versteegh and Dieks, namely a paper by Jaynes [28] who makes the point, following a similar analysis by Pauli, that the classical thermodynamic definition of entropy as the integration of d$Q/T$/ over a reversible path is only introduced in the context of constant particle number. This means that there is always (ultimately, see Ehrenfest and Trkal [29]) an arbitrary integration function (not an integration constant, but a function of $N$) that can be used to yield the desired extensivity of the entropy.

allowed state. Strictly speaking, this means that we can't talk, for instance, about an electron on Earth without mentioning all the electrons on the Moon in the same breath.

Why might Fleischhauer say this?[5] The answer is, from one perspective at least, rather straightforward. The universal superposition to which Fleischhauer refers arises because in reality, we never have perfect confinement of particles: there is no such thing as the infinite potential well beloved of introductory quantum physics courses, and there is therefore some finite (albeit extremely small) probability for tunnelling. Thus, in this sense, an electron on the Earth is indeed indistinguishable from an electron on the Moon (or on Alpha Centauri).

But what really matters, of course, are the effects that this type of "coupling" might have on experimental measurements. And for electrons separated by centimetres, let alone light years, those effects are, to put it mildly, utterly negligible. If we consider a "double-well" system for an electron on Earth and an electron on Alpha Centauri, the energy-level splitting is unimaginably tiny (and beyond anything, we could ever begin to hope to measure), and the timescale for evolution of the quantum state exceeds the age of the universe.

So *in any practical sense*, position can indeed be used to distinguish quantum particles. This is why we can treat electrons in well-separated atoms as being distinguishable. In principle, the electrons are indeed described by a single multiparticle ("universal") wave function and are thus indistinguishable. In practice, however, the spatial extent of the particle wave packet is such that we can treat electrons in atoms separated by distances much greater than their equilibrium bond length as distinguishable. Only when those atoms are brought together so that there is appreciable overlap of electronic wave functions, as in chemical bond formation or, as we shall discuss below, a DFM experiment can one state that the electrons on each atom become indistinguishable.

Following this lengthy "detour" on the topic of distinguishability versus indistinguishability, we are now finally at the point where we can return to a consideration of that back-of-an-envelope argument for the PEP which was mentioned above.

## 3 Statistics, Symmetry and Spin

Let us take a system where identical quantum particles cannot be distinguished from another. As the particles are indistinguishable then when we compute the probability density for the system, i.e. $|\Psi|^2$, we must get the same answer regardless

[5]It is perhaps worth noting at this point that the "interconnectedness" to which Fleischhauer alludes in this quote, and its relevance (or not) to the Pauli exclusion principle, was the subject of a great deal of sometimes ill-tempered online debate following the BBC's broadcast of a popular science lecture on quantum mechanics by Brian Cox, which included a discussion of the PEP. Jon Butterworth's post for The Guardian [31] is a short, clear and entertaining discussion of the furore and the physics surrounding Cox's lecture.

of how we arrange the particles, i.e. their spatial positions have no influence on the probability density. We will consider a very simple system with just two particles whose positions are $\mathbf{r}_1$ and $\mathbf{r}_2$ and whose single particle wave functions are $\psi_1$ and $\psi_2$, respectively. If we cannot distinguish Particle 1 from Particle 2, then it is clear that

$$|\Psi(\mathbf{r}_1, \mathbf{r}_2)|^2 = |\Psi(\mathbf{r}_2, \mathbf{r}_1)|^2 \tag{1}$$

This means one of two things. Either

$$\Psi(\mathbf{r}_1, \mathbf{r}_2) = \Psi(\mathbf{r}_2, \mathbf{r}_1) \tag{2}$$

or

$$\Psi(\mathbf{r}_1, \mathbf{r}_2) = -\Psi(\mathbf{r}_2, \mathbf{r}_1) \tag{3}$$

To meet the condition imposed by Eq. 2, we must have the following two-particle wave function:

$$\Psi(\mathbf{r}_1, \mathbf{r}_2) = \frac{1}{\sqrt{2}}(\psi_1(\mathbf{r}_1)\psi_2(\mathbf{r}_2) + \psi_2(\mathbf{r}_1)\psi_1(\mathbf{r}_2)) \tag{4}$$

Or to satisfy Eq. 3, we need the following:

$$\Psi(\mathbf{r}_1, \mathbf{r}_2) = \frac{1}{\sqrt{2}}(\psi_1(\mathbf{r}_1)\psi_2(\mathbf{r}_2) - \psi_2(\mathbf{r}_1)\psi_1(\mathbf{r}_2)) \tag{5}$$

Equation 4 represents what is called the symmetric case, while Eq. 5 is termed the antisymmetric case.[6] The antisymmetric equation leads us to a simple, but exceptionally important, result—a result that is at the very core of how the universe behaves because it is ultimately responsible for the stability of matter [32–34]. Note what happens when we make $\psi_1 = \psi_2$ in Eq. 5 (or, in other words, we put both particles in the same quantum state): *the two-particle wave function, $\Psi$, vanishes. This* is the essence of the PEP: in the antisymmetric case, no two particles can exist in the same quantum state.[7] (We should also stress that the exclusion principle is *not equivalent* to the statement that fermions have antisymmetric wave functions.

[6]The use of the terms symmetric and antisymmetric follows from Eq. 2 (where $\Psi$ is a symmetric function with respect to the exchange of coordinates) and Eq. 3 (where $\Psi$ is an antisymmetric function). Note also that the factor of $\frac{1}{\sqrt{2}}$ in Eqs. 4 and 5 arises from normalisation of the wave function.

[7]We are neglecting explicit consideration of the spin contribution here—see Sect. 3.1. Moreover, we are making drastic simplifications regarding the treatment of many electron systems in order to put across the "essence" of the exclusion principle. For example, Eqs. 4 and 5 are approximations because, in reality, there are many more contributing terms (as in the configuration interaction method of quantum chemistry. See Kantorovich [35] for a summary).

Rather, the exclusion principle follows from the antisymmetric character of fermions).

A rather remarkable observation is that *only* antisymmetric and symmetric wave functions are found in nature for fundamental particles, i.e. we only have bosons (symmetric state) and fermions (antisymmetry). No other particles have been found that fall outside these symmetry classes.[8] As Omar [24] points out in a comprehensive and very readable review of the ramifications of indistinguishability in quantum mechanics, this existence of only symmetric and antisymmetric states[9] is best described as a postulate (the "symmetrisation postulate"). And, disconcertingly, it is a postulate that apparently cannot be deduced from the framework of quantum mechanics (either the non-relativistic or relativistic "breeds" of the theory). In other words, we simply have to accept that only bosons and fermions exist (or, at least, we have no good experimental evidence to date for fundamental particles arising from other rather more exotic statistics/symmetries such as parastatistics, see Omar [24]). In this sense, we have progressed very little since Pauli voiced his misgivings about the origin of the exclusion principle almost seventy years ago:

> I was unable to give a logical reason for the Exclusion Principle or to deduce it from more general assumptions… in the beginning I hoped that the new quantum mechanics would also rigorously deduce the Exclusion Principle.

## *3.1 Putting a Spin on the Story*

All known fundamental particles are either bosons or fermions. (Within the standard model, fermions are "matter" particles, whereas bosons are generally force "carriers".[10] Again, we are not including quasiparticles in the discussion.). All bosons have integer spin, while fermions have half-integer spin. Clearly, there must be a strong connection between spin and symmetry. Indeed, this is known as the spin–statistics theorem and holds not just for individual particles but composites of fundamental particles.

This link between spin, statistics and the exclusion principle, however, very much appears not to be something that can be deduced from non-relativistic

---

[8]Note, however, that the key principle underlying the concept of *supersymmetry* is that bosons can be converted into fermions and vice versa. Supersymmetry therefore introduces a bosonic partner for every fermion (and, again, vice versa). To the chagrin of (some of) the particle physics community, however, any evidence for supersymmetry remains frustratingly elusive. Moreover, we are omitting any discussion of quasiparticles here. The results of measurements of two-dimensional systems exhibiting the fractional quantum Hall effect have been interpreted in terms of anyons [36] and quasiparticles with mixed symmetry.

[9]…for the *total* wave function. Again, see Sect. 3.1.

[10]…although the Higgs boson is an honourable exception.

quantum mechanics. This is the origin of the statement from Feynman quoted at the start of this chapter—the link between spin and the exclusion principle is "deep down" in relativistic quantum mechanics. More recently, Bartalucci et al. [37] have put it like this:

> Although the principle has been spectacularly confirmed by the number and accuracy of its predictions, its foundation lies deep in the structure of quantum field theory and has defied all attempts to produce a simple proof…

This means that within the non-relativistic quantum framework, the spin–statistics–symmetry link is generally accepted as a dictum, although alternative non-relativistic approaches have certainly been explored [21]. Duck and Sudarshan [38] detail a proof of the spin–statistics theorem which can be "recast" in non-relativistic quantum field theory, but only if an aspect of their proof which stems from relativistic quantum theory (via Lorentz invariance) can be invoked as a postulate.

Notwithstanding its essential relativistic origin, the spin contribution can be incorporated into the particle wave function in non-relativistic quantum mechanics in a straightforward fashion via the introduction of the spin orbital. A spin orbital is a product of a spatial wave function (such as those described in the preceding section) and a spin function, which we can represent as $\chi(\uparrow)$ or $\chi(\downarrow)$ for the spin-up and spin-down states, respectively. So, if we use $x$ as a variable which incorporates both the spatial and spin coordinates, and we switch to using $\phi$ to represent only the spatial part (so that we can, as per convention, use $\psi$ to represent the wave function), we have the following for the spin-up state of an electron:

$$\psi(\mathbf{x}_1) = \phi(\mathbf{r}_1)\chi(\uparrow) \tag{6}$$

We therefore now have two options for ensuring antisymmetry in a two-electron (or multi-electron) system: either the spatial part or the spin part can lead to an antisymmetric total wave function, $\Psi(\mathbf{x}_1, \mathbf{x}_2)$. In other words, if two electrons have opposite spin states, then there is no constraint on the spatial wave function. But this is nothing more than the statement of the PEP given earlier: no two electrons can exist in the same quantum state.

## 4 The Origin of Pauli Repulsion: A Gedankenexperiment

At short interatomic or intermolecular separations, Pauli repulsion[11] is much stronger than any electrostatic interaction, increasing very rapidly with decreasing distance between atoms or molecules. Recall, for example, that the Pauli repulsion term in the Lennard–Jones potential is modelled not with a $\frac{1}{r}$ dependence, as one would expect for a classical electrostatic interaction (between point charges), but

[11] We focus throughout this chapter only on fermions. For bosons, and as discussed by Mullin and Blaylock [39], an effective *attractive* force is often invoked.

with a $\frac{1}{r^{12}}$ function. This $\frac{1}{r^{12}}$ dependence is, of course, purely empirical in the Lennard–Jones (L-J) potential—it has no grounding in theory—but, nonetheless, the exceptionally high sensitivity of the repulsive interaction to small changes in interatomic/intermolecular separation is captured well by the functional form.

Of course, and as Baerends [40] discusses in a clear overview of Pauli repulsion effects in adsorption, we are dealing not with point charges and a pure Coulombic interaction but with a screened Coulomb potential and delocalised electron "clouds". The overlap of the electron clouds at short separations leads in a classical model, and perhaps counter-intuitively, to an *attractive* electrostatic interaction. It is only when the interatomic separation becomes so small that nuclear repulsion dominates that the overall electrostatic force becomes repulsive.

Thus, and as we hope is abundantly clear from previous sections, we cannot expect to understand electron repulsion due to Pauli exclusion in the context of classical electrostatics. The fundamental origin of the repulsion comes from, as we have seen, the physical impossibility of "squeezing" two fermions into the same quantum state. But the central question is this: just how does the exclusion principle translate into a physically measurable interaction? We will see in the following section how DFM allows us to directly probe the exclusion-derived repulsion between the electron density of two atoms or molecules. Before we consider the results of the real-world experiment, however, it is very helpful to think about a "stripped-down" system involving the overlap of two single particle wave functions (see Sect. 3) [41–43]. This "*Gedankenexperiment*", if you will, provides compelling insights into the origin of Pauli repulsion.

First, recall that the kinetic energy operator is $-\frac{\hbar^2}{2m}\nabla^2$. The curvature of a wave function therefore determines its kinetic energy (via the Laplacian, $\nabla^2$). Wilson and Goddard's approach [41] to elucidating the origin of Pauli repulsion was to compare the kinetic energy (KE) of a Hartree product of the wave functions for two same-spin electrons with the KE of an antisymmetrised product (see Fig. 2). A Hartree product is simply the following:

$$\Psi_{\text{Hart}}(\mathbf{r}_1, \mathbf{r}_2) = \psi(\mathbf{r}_1)\psi(\mathbf{r}_2) \tag{7}$$

As should be clear from Sect. 3, the multiparticle wave function $\Psi_{\text{Hart}}$ is not antisymmetric (nor does it take into account indistinguishability of the particles) and is therefore in general not appropriate to use to describe fermions. However, we can take the Hartree product as a representation of the system when the PEP is "suppressed" and determine the resulting kinetic energy.

In order to incorporate Pauli exclusion, we have to consider a multiparticle wave function which is appropriately *antisymmetrised.* Slater introduced an elegant method of enforcing this antisymmetry requirement via the determinant approach which now bears his name [44]. Wilson and Goddard [41] focussed on the orthogonality of orbitals which is generally *imposed* in approaches which treat the

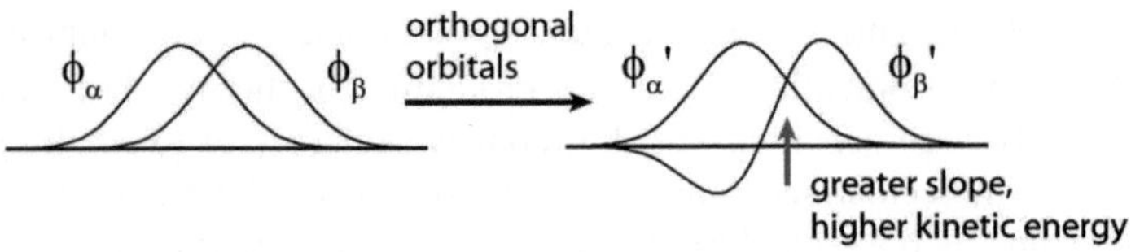


**Fig. 2** The effective repulsion due to Pauli exclusion stems from the change in the curvature of the wave function due to the requirement for antisymmetrisation in fermion systems. One approach to visualising this is to consider the orthogonalisation of orbitals (which is placed as a constraint on Slater determinant approaches to constructing a multiparticle wave function). Higher wave function curvature leads to a higher kinetic energy. Equivalently, higher curvature is accounted for in Fourier space by higher spatial frequency (momentum) components. Figure taken from the PhD thesis of Julian Su [45]. © Julian Su (2007)

multiparticle wave function in terms of (a sum of) Slater determinants (see Fig. 2, taken from the PhD thesis of Julius Su [45]). We note, however, that orthogonality is a constraint on the multiparticle wave function *that is not strictly necessary* [35] and, as discussed by Beylkin et al. [46], leads to ever-increasing levels of computational expense as the size of a system grows.

Nonetheless, to ensure antisymmetry (i.e. the requirement of Eq. 5), wave function slope and curvature must necessarily increase and thus, the overall picture emerging from Fig. 2 is correct (even if one does not invoke orthogonality as the root cause of the increase in wave function curvature). This change in curvature results in a corresponding increase in kinetic energy. A complementary explanation from a Fourier analysis perspective, as noted in the following section, is that the increase in curvature of the wave function necessitates the introduction of higher spatial frequency contributions, i.e. higher *momentum* components). It is this increase in KE (or momentum) which is responsible for the majority of Pauli repulsion.

There are two important assumptions built into this description of Pauli exclusion, however. First, we have adopted a "pairwise" approach to considering electron–electron interactions when, in reality, Pauli exclusion is an *n*-body, rather than a two-body problem. The second, and related, issue is that the modification of the wave function due to orthogonalisation will mean that the electron density will be distributed differently, affecting electron–electron interactions and giving rise to the effect known as correlation. Interactions between same-spin electrons go by the name *Fermi correlation*, whereas those between opposite-spin electrons are known as *Coulomb correlation.*[12] Nonetheless, the dominating contribution to Pauli repulsion is the pure quantum mechanical component arising from wave function antisymmetry.

[12]The combined contributions of the exclusion principle and electron correlation produce the exchange-correlation contribution to the functional in density functional theory.

## 5 Is There a Pauli Exclusion *Force*?

Having spent much of the chapter up to this point using the term "Pauli repulsion", it might seem a little perverse for us to now pose the question as to whether there is a Pauli exclusion force or not (particularly as the experimental technique we are considering is DFM). Notwithstanding the use of "*Pauli repulsion*" or "*Pauli exclusion force*" in the DFM literature—and, more broadly, throughout very many areas of science (spanning, for example, particle physics, single-molecule imaging and spectroscopy, astrophysics[13] and cosmology)—a number of authors have made the claim that Pauli exclusion does not produce a force in the traditional sense. Mullin and Blaylock [39], in particular, present a set of arguments as to why they are of the opinion that couching the effects of Pauli exclusion in terms of a repulsive *force*, or exchange *force*, can be rather misleading. Indeed, they go so far as to argue—and we quote directly from their paper—that *"there is no real force due to Fermi/Bose symmetries"*, citing, amongst others, Griffiths' description of the effects of Pauli exclusion [48]:

> We call it an exchange force but it is not really a force at all—no physical agency is pushing on the particles; rather it is purely a geometric consequence of the symmetrization requirement.

What does Griffiths (and, by extension, Mullin and Blaylock) mean by this?

To back up their assertion that Pauli "repulsion" is not a force in the traditional sense, Mullin and Blaylock consider a number of "archetypal" physicochemical phenomena where the exclusion principle plays a key role. Arguably, the most instructive of these is their discussion of the changes in momentum in a classical gas as compared to a Fermi gas. We encourage the reader to follow the detail of the analysis in Section II of their paper (under the subsection entitled *Virial Expansion*) and restrict ourselves here simply to highlighting the central point they make.

Consider first a classical ideal gas in a container. Pressure, $P$, arises from the combined impacts of each atom of that gas on the walls of the container and is given by the force per unit area. Force, in turn, is the rate of change of momentum. The mean force, $\bar{F}$, which each individual molecule of the gas contributes, is $\bar{F} = \Delta p/\Delta t$, where $\Delta p$ is the momentum change on striking the wall. (This is twice the atomic momentum because the sign of the momentum flips on collision). $\Delta t$ is the time required for an atom to cross the container, i.e. $\Delta t = mL/\bar{p}$ where $L$ is the width of the container and $m$ is the atomic mass. The key point in the classical case is this: if we make the volume of an ideal gas smaller or we introduce repulsive interactions (with no change in temperature), the pressure of the gas will rise because of a decreased $\Delta t$ due to a change in (the effective) $L$ arising from collisions, but $\bar{p}$ *remains the same*. (Recall that for a classical gas, the root-mean-square momentum, $p_{\text{rms}}$ is $\sqrt{3mk_BT}$).

[13]The Pauli exclusion principle prevents the collapse of white dwarf and neutron stars. See [47].

Compare this to what happens for a Fermi gas subject to the exclusion principle. The effect of the exclusion principle is to modify the *momentum distribution*. Mullin and Blaylock argue that this is subtly different to what happens for the classical gas when repulsive interactions are introduced. Classically, the repulsive forces raise the pressure of the gas because the collisions and deflections of the atoms change the atomic transit time. Quantum mechanically, the momentum distribution is "intrinsically" modified because of the higher curvature of the wave function which results from the exclusion principle. Position and momentum are conjugate variables and are thus two sides of the same coin—Fourier transformation allows us to switch between the two (entirely equivalent) representations. The higher wave function curvature demanded by Pauli exclusion is entirely equivalent to stating that higher spatial frequency components are required in reciprocal (i.e. momentum) space.[14] It is this intrinsic symmetry-driven modification of the momentum distribution which raises the pressure of the Fermi gas.

It is worth lifting another couple of quotes from Mullin and Blaylock's paper to highlight just how strongly opposed they are to equating Pauli exclusion with a repulsive force:

> The idea of an effective repulsion between fermions ignores the real physics and gives a very poor analogy with classical repulsive gases…we offer the following guiding principle regarding statistical symmetries: "May the force be *not* with you".

Is this degree of antiforce scepticism justified, however?

## 6 Beyond Gedanken: Exploiting Exclusion in Force Microscopy

At this point, the pragmatic scanning probe microscopist could quite reasonably take issue with the preceding arguments because the primary experimental observable in a DFM experiment is the frequency shift of the probe. And this, via the Sader–Jarvis formalism [49], for example, can be converted directly to a tip–sample *force*. The effects of Pauli exclusion are directly measurable in DFM because they shift the resonant frequency of the probe–cantilever system, and this ultimately can be interpreted as a change in the tip–sample force. Notwithstanding the arguments put forward by Mullin and Blaylock [39], and Griffiths [48], amongst others, if Pauli exclusion is not giving rise to a force, then it certainly very much looks like it in a DFM experiment.

The resolution of this apparent conflict may lie, as Moll et al. have discussed in a recent paper focussed on the interpretation of submolecular resolution DFM images [50], in the virial theorem. Slater showed in the 1930s that the virial theorem can be applied to a molecule [51], assuming that the nuclei are fixed in place by external

[14]This, of course, is the fundamental origin of the Heisenberg uncertainty principle.

forces. The total electron energy, $E$, is related to the electronic kinetic energy, $T$, and potential energy, $V$, as follows:

$$T = -E - r\left(\frac{\mathrm{d}E}{\mathrm{d}r}\right) \tag{8}$$

$$V = 2E + r\left(\frac{\mathrm{d}E}{\mathrm{d}r}\right) \tag{9}$$

The electronic kinetic energy and potential energy are thus coupled via the virial theorem. Moll et al. [50] claim that, despite the Pauli exclusion force being non-conservative in character, if it is assumed that we have a diatomic (or dimolecular) system with one degree of freedom—as is the case for the tip–sample system in DFM—the Pauli energy and the increase in electronic kinetic energy can be related as follows:

$$E_{\mathrm{Pauli}}(z) = \frac{1}{z}\int_{z}^{\infty} \Delta E_{kin}(z')\mathrm{d}z' \tag{10}$$

where $z$ is the interatomic/intermolecular separation. The issue of extracting accurate measures of non-conservative forces from the frequency shift observable in DFM, however, continues to attract considerable debate and discussion. For example, the Sader–Jarvis inversion technique [49] widely applied to extract forces from frequency shift versus separation curves must, as John Sader and his co-authors themselves highlight [52], be applied with great care under conditions were there is a significant contribution from non-conservative forces.

Although the authors cited in the previous section propose reasons for drawing a distinction between a traditional force and the effects arising from Pauli exclusion, the increase in kinetic energy and momentum resulting from the requirement for wave function antisymmetry nonetheless ultimately result in an interaction which is measured as a repulsive force in a DFM experiment. That is, the connection between the change in kinetic energy and the total energy of the tip–sample system appears to result in a measurable, and positive (i.e. repulsive), contribution to the frequency shift due to the PEP. What is important to realise from the previous sections, however, is that Pauli exclusion really is not comparable to other types of interparticle interaction. In this sense, it is a phenomenon which is distinct from the four fundamental forces, i.e. strong, weak, electromagnetic (in particular) and, if the graviton exists, gravity.

## *6.1 Intramolecular Imaging*

Although DFM's "sibling" technique, scanning tunnelling microscopy (STM), has long been capable of submolecular resolution imaging, in the sense that molecular orbital density can be probed (see an earlier volume of this Springer series on Atom and Single Molecule Machines [53]), only DFM is capable of resolving the chemical framework or atomic structure of a molecule. This is because STM probes orbital density only within a specific energy window (set by the potential difference between the tip and sample), and in conventional tunnelling microscopy, therefore, only the frontier molecular orbitals are accessible.[15] The spatial distribution of the frontier orbital density generally does not map onto the atomic positions and indeed often bears very little relationship to the "ball-and-stick" models of molecules so familiar to chemists and physicists.

As Giessibl has highlighted [54], however, DFM is not restricted to probing the frontier orbital density and is sensitive to the total charge density. This is because intramolecular forces depend on the total electron density, rather than the density of states within a certain energy window [55]. The sensitivity of DFM to the total electron density is particularly pronounced when in the Pauli exclusion regime of imaging, i.e. at very small tip–sample separations. Figure 1 at the start of this chapter shows very clearly that, unlike STM, DFM in this Pauli exclusion regime produces images which are remarkably similar to the ball-and-stick structural models of molecules.

On the basis of Fig. 3 (and related theoretical and experimental data), Moll et al. [50] argue that there is a close connection between the charge density of a molecule and the increase in electron kinetic energy due to Pauli exclusion. This assumes that (a) the arguments regarding wave function curvature outlined in Sects. 4 and 5 provide an accurate model of electron–electron interactions at the tip–sample junction, and (b) the dominant effect is the change in kinetic energy and that this can be "deconvolved" from the overall response of the electron density as a function of the tip–sample separation. They approximate the complicated relationship between the increase in kinetic energy and the separation of two atoms with different nuclear charges (see Eq. 6 of their paper) as follows:

$$\Delta E_{\text{kin}}(z) = A\rho_s(z)^B \qquad (11)$$

where $z$ is the interatomic/intermolecular separation, $\rho(z)$ is the sample charge density at separation $z$, and $A$ and $B$ are two tunable parameters. As can be seen in Fig. 3, this simple power law model, which involves no explicit consideration of the probe, provides good agreement with experimental frequency shift images of a 3,4,9,10-perylenetetracarboxylic dianhydride (PTCDA) molecule. We also include in Fig. 3, again from Moll et al.'s paper, a comparison of the charge density of the PTCDA molecule with the increase in kinetic energy calculated using the simple model of Eq. 11. There is again apparently good agreement, adding support to the idea that DFM is sensitive to the total charge density of the system.

[15] In the scanning tunnelling hydrogen microscopy (STHM) [9–11] variant of STM mentioned earlier, this constraint can be circumvented.

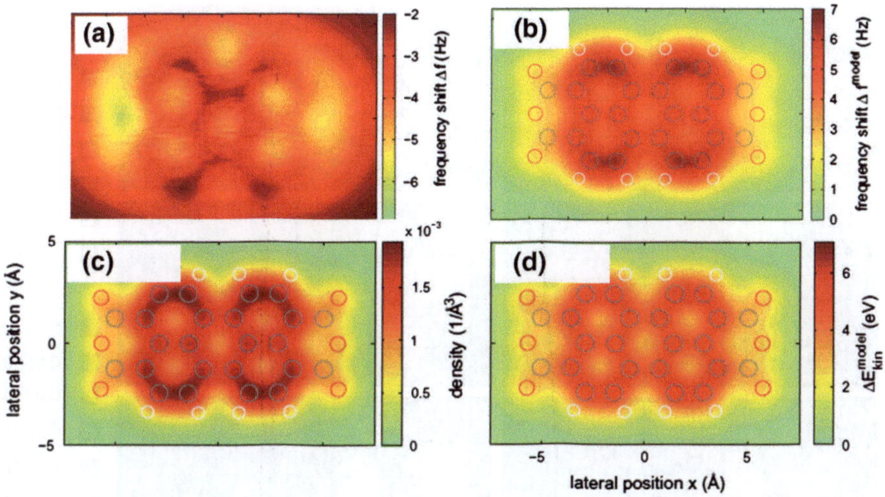


**Fig. 3** Comparison of **a** experimental frequency shift image and **b** a simulated frequency shift image for a 3,4,9,10-perylenetetracarboxylic dianhydride (PTCDA) molecule calculated on the basis of the Pauli exclusion-derived change in electron kinetic energy. **c** Charge density of a PTCDA molecule at a given tip–sample separation. Compare with **d** the change in kinetic energy at the same tip–sample separation. Figure adapted from Moll et al. [50]. © Institute of Physics Publishing (2012)

What is not included in the model used to generate the simulated images in Fig. 3—although Moll and co-workers deal with this point elsewhere [56]—is the relaxation or bending of the CO molecule at the tip apex as it is moved across the underlying PTCDA molecule. It turns out that this is an extremely important contribution to the observation of intramolecular and intermolecular contrast in DFM images and we will return to it in the final section.

## 6.2 Density Depletion

The modification of the curvature and spatial extent of the tip–sample wave function due to Pauli exclusion produce extensive modification to the total electron density of the system. A key aspect of this is the generation of regions of density depletion. Baerends [40] discusses the importance of density depletion in the context of the Ag–O bond where a substantial degree of Pauli exclusion-derived depletion around the centre of the bond is observed.

As a more recent example in the context of DFM, a number of the authors of this chapter have explored the importance of density depletion in the interpretation of images taken in the Pauli exclusion regime. The molecular system we used is that shown in Fig. 1g—a hydrogen-bonded assembly of naphthalenetetracarboxylic diimide (NTCDI) molecules on a passivated silicon surface. Figure 4 shows a

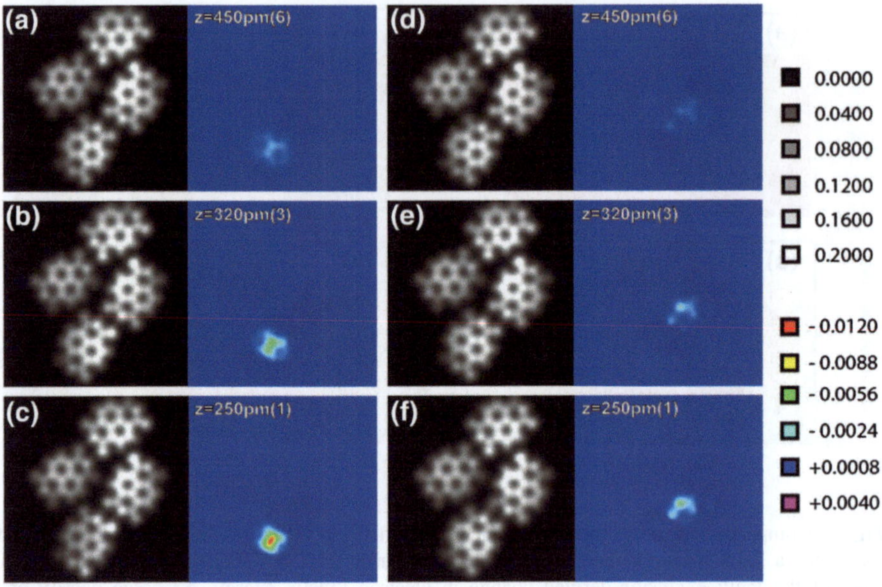


**Fig. 4** Total electronic density (TED) and electron density difference (EDD) calculated for an NTCDI assembly plotted 100 pm above the molecular plane for a variety of different tip heights. At each tip height in a simulated $F(z)$ curve, the EDD was obtained by first calculating the TED for (i) the isolated surface and (ii) the isolated NTCDI tip. These two densities were then summed together and subtracted from the relaxed total density for the full system. The remaining quantity is the EDD. This quantifies the fraction of charge which is redistributed due to the interaction of the DFM tip and the NTCDI molecule. The TED (*left*) and EDD (*right*) are shown for an oxygen-down NTCDI probe molecule at **a–c** the C–C location on an NTCDI molecule, and **d–f** at the intermolecular H-bond location for the different tip heights specified in each figure. Figure from Sweetman et al. [4]. © Nature Publishing Group (2014)

comparison of the total electron density for an NTCDI assembly versus the density difference at a number of different $z$ positions of the tip above a C–C bond (Fig. 4a–c) and above an intermolecular region where hydrogen bonding is expected [4]. Pauli exclusion results in strong tip-induced electron depletion above both the intermolecular and intramolecular bond regions.

The most important insight to be derived from this analysis of density depletion is that, as is always the case in any type of scanning probe experiment (and as is well-understood across the SPM community), the influence of the tip on the imaging process must *always* be carefully considered. Tip–sample interactions and convolution have been an issue for STM since its invention, of course, but with the advent of DFM imaging in the so-called Pauli regime, the probe can certainly no longer be treated as just a perturbation of the electronic structure. The tip–sample separation for the type of high-resolution images shown in Fig. 1 is such that the repulsive Pauli component makes a strong contribution—the tip interacts heavily with the underlying molecule adsorbed on the sample surface. In this sense, the sample–tip apex system should be considered as one large molecule.

In the following, and final, section of this chapter, we will see just how important a role the tip can play in generating high-resolution DFM images.

## 7 But Do We *Really* See Bonds?

A key "ingredient" in attaining intramolecular contrast in DFM is the passivation of the tip apex. Gross et al. [1] first showed that CO was a particularly appropriate molecule to use for imaging submolecular structure. (In the same paper, and in subsequent work [57], they compared the imaging capabilities of CO with those of other species adsorbed at the tip apex). Although deliberate functionalisation with CO is certainly not necessary to obtain intra- (and inter-)molecular contrast [4], carbon monoxide remains the molecule of choice at present for high-resolution DFM.

It turns out that CO is very far from a rigid probe, however, and the tilting of the molecule at the tip apex plays an essential role in the imaging process. The flexibility of CO has been studied in some detail by a number of groups [1, 56, 58–60], but it is a very recent paper [5], available only at the condensed matter arXiv at the time of writing, on which we would like to focus here. This paper provides particularly telling insights into the extent to which the probe itself contributes to the structure seen in molecular and submolecular images.

Hapala et al. [5] use an exceptionally simple, but remarkably powerful, model to simulate DFM (and STHM [9–11]) images acquired either with a CO probe or with any other type of tip apex. They represent the tip–sample geometry as shown in Fig. 5 and account for interactions between the probe and sample molecule using analytical Lennard–Jones potentials. It is very important to note that *no account is taken of intra- or intermolecular charge density* in this model: the approach adopted by Hapala et al. uses only the coordinates of the atoms within the molecule under study—electron density due to bonding between those atoms is not incorporated in their model. In other words, the force field does not rely on the electron density of the system. Although this might at first glance appear to be a rather crude approach (as compared to, for example, modelling the system using an ab initio method such as density functional theory), it is nonetheless the case that their "stripped-down" model accurately reproduces the experimental data. This is the acid test of any theory or simulation.

Figure 6 shows a comparison between experimental DFM images and the output of Hapala et al.'s simulations for two systems comprising assemblies of 8-hydroxyquinoline tetramers and NTCDI molecules (as discussed above in the context of Fig. 4), respectively. For both of these systems, intermolecular interactions are mediated by hydrogen bonding. Note, however, how the sharp intra- and intermolecular features in the simulated image of Fig. 6a agree extremely well with those in the experimental data shown in Fig. 6b, despite the absence of any intra- or

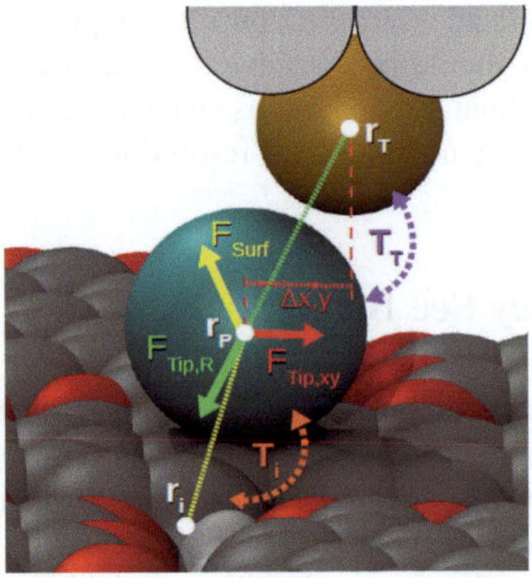


**Fig. 5** Schematic model of the tip–sample geometry used by Hapala et al. [5] in their analysis of the origin of intra- and intermolecular contrast in DFM images. The final metal atom at the tip apex and the "probe particle" are shaded in gold and cyan, respectively, with the underlying molecular layer represented by the standard space-filling model. The coloured vectors show the various forces on the tip: $F_{\mathrm{Tip},R}$ (*green*) is the radial force; $F_{\mathrm{Tip},xy}$ (*red*) is the lateral force; and $F_{\mathrm{Surf}}$ (*yellow*) is the force due to the sample molecules. ($T_i$ and $T_t$ refer to tunnelling processes not of interest in this chapter.) Taken from Hapala et al. [5]

intermolecular charge density in the model. Figure 6c, d similarly shows a comparison between the "flexible tip" model and DFM images of a hydrogen-bonded NTCDI assembly [4] taken by a number of the authors of this chapter. Again, intramolecular and intermolecular features are observed in the simulated image, despite the absence of any charge density due to covalent or hydrogen bonding.

It therefore would appear that the flexibility of the probe molecule plays a major role in the imaging of intra- and intermolecular structure. But we have seen in previous sections that there is also a close correspondence between images simulated on the basis of an increase in electron kinetic energy due to Pauli exclusion and the experimental frequency shift data [50]. Moreover, the intensity of intramolecular bonds as observed by DFM is related to the Pauling bond order [56], i.e. the charge density. Similarly, the DFM images of de Oteyza et al. [2] clearly show a pronounced difference between single, double and triple bonds. The key issue is therefore the extent to which the response of the tip to interatomic and/or intermolecular charge density is a "first-order" versus "second-order" contribution to the imaging mechanism, as compared to the flexibility of the probe. This is currently a very active area of debate.

In order to explore the influence of tip relaxation on the DFM images of NTCDI shown in Figs. 4 and 6, we (i.e. Sweetman et al. [4]) generated simulated images using a variant of DFT where both the atomic geometry and the electronic structure of the system were "frozen". Despite the lack of probe relaxation, a weak feature at the expected position of the hydrogen bond was observed. Nonetheless, another question remains: to what extent might convolution of the tip's electron density

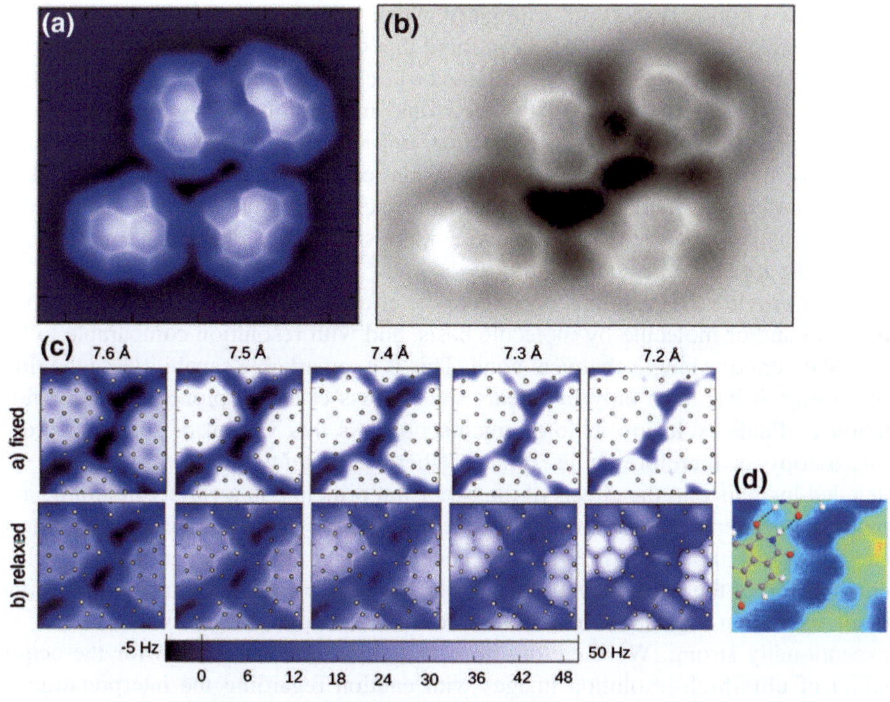


**Fig. 6** **a**, **b** Comparison of a simulated DFM image of a hydrogen-bonded assembly of 8-hydroxyquinoline molecules (from Hapala et al. [5]) with the corresponding experimental DFM image taken from Zhang et al. [3]. **c** Series of simulated frequency shift images at different tip–sample separations, again from Hapala et al. [5], of NTCDI molecules using a (*top row*) unrelaxed, and (*bottom row*) relaxed tip. **d** Experimental frequency shift image for comparison (from Sweetman et al. [5])

with molecular charge density at the edge of a molecule account for the observation of "intermolecular" features? In the supplementary information file associated with their paper, Hapala et al. [5] suggest that this convolution effect could be as strong as the interaction of the probe with any charge density due to an intermolecular bond. This is an exceptionally important issue which needs to be addressed in a timely fashion by the scanning probe microscopy community.

## 8 Conclusions

The history of the development of the PEP provides fascinating insights into just how problematic it is to associate purely quantum mechanical concepts with classical "real-world" analogies. In this sense, it is a shame that Pauli's *Zweideutigkeit* term did not gain wider acceptance as it is a less misleading, albeit rather more prosaic, description than "spin". Similarly, when we describe the PEP as giving rise to a repulsive force, we should bear in mind that the origin of the repulsion detected in DFM is not at all adequately explained via classical analogies. The interaction arises from the modification of the electrons' momentum distribution due to the increased curvature of the wave function imposed by the requirement for anti-symmetrisation in fermion systems. Classical analogies will clearly fail. Understanding the fundamental origin of the increased wave function curvature is ultimate, as Feynman puts it in the quote at the start of this chapter, "*deep down in relativistic quantum mechanics*".

DFM provides us with direct access to the effects of Pauli exclusion on an atom-by-atom and/or molecule-by-molecule basis, and with resolution comparable to the spatial extent of a single chemical bond. This is a remarkable capability. At the time of writing, it has been only five years since Gross et al. [1] pioneered the exploitation of Pauli exclusion in force microscopy. As this variant of scanning probe microscopy is therefore in its infancy, there is potentially immense scope for detailed insights into the effects of the exclusion principle in a variety of atomic and molecular systems. However, every probe microscope image—indeed, every image (regardless of the technique used to generate that image)—is, at some level, a convolution of the properties of the sample and those of the imaging system. In the Pauli exclusion regime of dynamic force microscopy, this convolution can be exceptionally strong. We therefore need to temper our enthusiasm for the acquisition of ultra-high-resolution images with caution regarding the interpretation of the data, as the examples included in this chapter clearly show.

**Acknowledgments** We are very grateful for financial support from the UK Engineering and Physical Sciences Research Council in the form of a fellowship (EP/G007837), from the Leverhulme Trust (through grant F/00114/BI) and from the European Commission's ICT-FET programme via the Atomic Scale and Single Molecule Logic gate Technologies (AtMol) project (www.atmol.eu), Contract No. 270028 and the ACRITAS Marie Curie Initial Training Network (www.acritas.eu). We are also very grateful for the support of the University of Nottingham High Performance Computing Facility. PJM thanks Christian Joachim (CNRS Toulouse) and Leo Gross (IBM Zurich) for helpful discussions and e-mail exchanges on the role of the exclusion principle in probe microscopy.

## References

1. Gross, L., Mohn, F., Moll, N., Liljeroth, P., Meyer, G.: Science **325**, 1110 (2009)
2. de Oteyza, D.G., Gorman, P., Chen, Y.-C., Wickenburg, S., Riss, A., Mowbray, D.J., Etkin, G., Pedramrazi, Z., Tsai, H.-Z., Rubio, A., Crommie, M.F., Fischer, F.R.: Science **340**, 1434 (2013)
3. Zhang, J., Chen, P., Yuan, B., Ji, W., Cheng, Z., Qiu, X.: Science **342**, 611 (2013)
4. Sweetman, A., Jarvis, S.P., Sang, H., Lekkas, I., Rahe, P., Wang, Y., Wang, J., Champness, N., Kantorovich, L., Moriarty, P.: Nat. Comm. **5**, 3931 (2014)
5. Hapala, P., Kichin, G., Wagner, C.F., Tautz, S., Temirov, R., Jelinek, P.: Phys. Rev. B **90**, 085421 (2014)

6. Pavlicek, N., Fleury, B., Neu, M., Niedenfuehr, J., Herranz-Lancho, C., Ruben, M., Repp, J.: Phys. Rev. Lett. **108**, 086101 (2012)
7. Pawlak, R., Kawai, S., Fremy, S., Glatzel, T., Meyer, E.: ACS Nano **5**, 6349 (2011)
8. Riss, A., Wickenburg, S., Gorman, P., Tan, L.Z., Tsai, H.-Z., de Oteyza, D.G., Chen, Y.-C., Bradley, A.J., Ugeda, M.M., Etkin, G., Louie, S.G., Fischer, F.R., Crommie, M.F.: Nano Lett. **14**, 2251 (2014)
9. Temirov, R., Soubatch, S., Neucheva, O., Lassise, A.C., Tautz, F.S.: New J. Phys. **10**, 053012 (2008)
10. Weiss, C., Wagner, C., Kleimann, C., Rohlfing, M., Tautz, F.S., Temirov, R.: Phys. Rev. Lett. **105**, 086103 (2010)
11. Weiss, C., Wagner, C., Temirov, R., Tautz, F.S.: J. Am. Chem. Soc. **132**, 11864 (2010)
12. Kichin, G., Weiss, C., Wagner, C., Tautz, F.S., Temirov, R.: J. Am. Chem. Soc. **133**, 16847 (2011)
13. Massimi, M.: Pauli's Exclusion Principle: The Origin And Validation Of A Scientific Principle. Cambridge University Press, Cambridge (2005)
14. Pauli, W.: Z. Angew. Phys. **31**, 373 (1925)
15. Pauli, W.: Z. Angew. Phys. **31**, 765 (1925)
16. Fermi, E.: Rendiconti Lincei **3**, 145 (1926)
17. Zannoni, A.: On the quantization of the monoatomic ideal gas. arXiv:cond-mat/9912229 [cond-mat.stat-mech]. Translation of Fermi's 1926 paper
18. Dirac, P.A.M.: Proc. Royal Soc. A **112**, 661 (1926)
19. Davies, P.C.W., Betts, D.S.: Quantum Mechanics. Stanley Thornes (Publishers) Ltd., Cheltenham, England (1994)
20. De Muynck, W.M.: Int. J. Theor. Phys. **14**, 327 (1975)
21. Berry, M.V., Robbins, J.M.: J. Phys. A **33**, L207 (2000)
22. Ginsberg, N.S., Garner, S.R., Hau, L.V.: Nature **445**, 623 (2007)
23. Fleischhauer, M.: Nature **445**, 605 (2007)
24. Omar, Y.: Contemp. Phys. **46**, 437 (2005)
25. Dieks, D., Versteegh, M.A.M.: Found. Phys. **38**, 923 (2008)
26. Dieks, D., Lubberdink, A.: Found. Phys. **41**, 1051–1064 (2011)
27. Versteegh, M.A.M., Dieks, D.: Am. J. Phys. **79**, 741 (2011)
28. Jaynes, E.,: The Gibbs Paradox. In Erickson, G.J., Smith, C.R., Neudorfer, P.O. (eds.): Maximum Entropy and Bayesian Methods. Springer, Berlin (1992)
29. Ehrenfest, P., Trkal, V.: Deduction of the Dissociation-Equilibrium from the Theory of Quanta and a Calculation of the Chemical Constant Based on This. In: Collected Scientific Papers, p. 414. North-Holland, Amsterdam (1959)
30. Rohlf, J.W.: Modern physics: from α to Z0. Wiley, New Jersey (1994)
31. Butterworth, J.: On Pauli and the Interconnectedness of all Things. http://www.theguardian.com/science/life-and-physics/2012/feb/28/1, (2012). [Online; accessed 07 July 2014]
32. Lieb, E.H., Thirring, W.E.: Phys. Rev. Lett. **35**, 687–689 (1975)
33. Lieb, E.H.: Rev. Mod. Phys. **48**, 553–569 (1976)
34. Lieb, E.H.: Bulletin Am. Math. Soc. **22**, 1–49 (1990)
35. Kantorovich, L.: Quantum Theory of The Solid State: An Introduction. Kluwer Academic Publishers, Dordrecht (2004)
36. Wilczek, F.: Phys. Rev. Lett. **49**, 957 (1982)
37. Bartalucci, S., Bragadireanu, M., Cargnelli, M., Catitti, M , Curceanu-Petrascu, C., Di Matteo, S., Egger, J.-P., Guaraldo, C., Iliescu, M., Ishiwatari, T., Laubenstein, M., Marton, J., Milotti, E., Pietreanu, D., Ponta, T., Sirghi, D.L., Sirghi, F., Sperandio, L., Widmann, E., Zmeskal, J. Phys. Lett. B. **641**, 18 (2006)
38. Duck, I., Sudarshan, E.G.C. Am. J. Phys. **66**, 284 (1998)
39. Mullin, W.J., Blaylock, G.: Am. J. Phys. **71**, 1223 (2003)
40. Baerends, E.J.: In Cluster models for surface and bulk phenomena **283**, 189–207 (1992)
41. Wilson, C.W., Goddard, W.A.: Theor. Chim. Acta **26**, 195 (1972)
42. Wilson, C.W., Goddard, W.A.: Theor. Chim. Acta **26**, 211 (1972)

43. Wilson, C.W., Goddard, W.A.: Chem. Phys. Lett. **5**, 45–49 (1970)
44. Slater, J.C.: Phys. Rev. **34**, 1293 (1929)
45. Su, J.: An electron force field for simulating large scale excited electron dynamics. http://thesis.library.caltech.edu/1598/, (2007). [Online; accessed 07 July 2014]
46. Beylkin, G., Mohlenkamp, M.J., Pérez, F.: J. Math. Phys. **49**, 032107 (2008)
47. Hänsel, P., Potekhin, A.Y., Yakovlev, D.G.: Neutron Stars 1: Equation of State and Structure. Springer, New York (2007) http://www.springer.com/gp/book/9780387335438
48. Griffiths, D.J.: Introduction to Quantum Mechanics. Prentice-Hall, Englewood Cliffs
49. Sader, J.E., Jarvis, S.P.: Accurate formulas for interaction force and energy in frequency modulation force spectroscopy. Appl. Phys. Lett. **840**(10), 1801–1803 (2004)
50. Moll, N., Gross, L., Mohn, F., Curioni, A., Meyer, G.: New J. Phys. **14**, 083023 (2012)
51. Slater, J.C.: The virial and molecular structure. J. Chem. Phys. **1**, 687–691 (1993)
52. Sader, J.E., Uchihashi, T., Higgins, M.J., Farrell, A., Nakayama, Y., Jarvis, S.P.: Nanotechnology **16**, S94 (2005)
53. Moriarty, P.: Submolecular resolution imaging of $C_{60}$:From orbital density to bond order. In: Advances in Atom and Single Molecule Machines Vol III: Imaging and Manipulating Molecular Orbitals, p. 195. Springer, Berlin (2013)
54. Giessibl, F.J.: High-speed force sensor for force microscopy and profilometry utilizing a quartz tuning fork. Appl. Phys. Lett. **73**, 3956 (1998)
55. Feynman, R.P.: Forces in molecules. Phys. Rev. **56**, 340–343 (1939)
56. Gross, L., Mohn, F., Moll, N., Schuler, B., Criado, A., Guitin, E., Pea, D., Gourdon, A., Meyer, G.: Science **337**, 1326 (2012)
57. Mohn, F., Schuler, B., Gross, L., Meyer, G.: Appl. Phys. Lett. **102**, 073109 (2013)
58. Sun, Z., Boneschanscher, M.P., Swart, I., Vanmaekelbergh, D., Liljeroth, P.: Phys. Rev. Lett. **106**, 046104 (2011)
59. Welker, J., Giessibl, F.J.: Science **336**, 444 (2012)
60. Weymouth, A.J., Hofmann, T., Giessibl, F.J.: Science **343**, 1120 (2014)

# Mechanical and Electrical Properties of Single Molecules

**Thilo Glatzel**

**Abstract** Molecular systems and devices are an strongly emerging technology. Various devices are already available, while others are still under development. The usage of molecules offers a great advantage to Si-based electronics, the functional groups of the molecules are adoptable allowing to artificially generate a huge variety of different properties and functionalities. However, the fundamental requirements of molecular engineering are not fully developed yet, mainly well-known molecules with limited functionality are used nowadays. To analyze and create new molecular structures providing, for example, a very high extinction coefficient while maintaining the chemical structure under environmental conditions would allow to create new types of solar cell conversion devices. This chapter introduces and reviews the research activities of the Nanolino-group at the University of Basel in the active field of the usage of scanning probe microscopy techniques to characterize the mechanical and electrical properties of molecular assemblies down to single molecules.

**Keywords** Scanning probe microscopy · Single molecules · Molecular assemblies · Insulators · KBr · NaCl

## 1 Introduction

Molecular electronics is based on the ultimate goal to use single molecules in electronic devices [1–3]. A major challenge is to mechanically stabilize and electrically decouple the molecules at and from a surface, respectively. Unfortunately, these requirements are somehow contradictory, while most molecular assemblies and single-molecular studies are done on metallic surfaces by scanning tunneling microscopy (STM) [4–6], the electrically decoupling requires insulating films or

T. Glatzel (✉)
Department of Physics, University of Basel, Klingelbergstr. 82, 4056 Basel, Switzerland
e-mail: thilo.glatzel@unibas.ch

 
P. Moriarty and S. Gauthier (eds.), *Imaging and Manipulation of Adsorbates Using Dynamic Force Microscopy*, Advances in Atom and Single Molecule Machines,
DOI 10.1007/978-3-319-17401-3_2

substrates. Controlled growth procedures of single molecules on insulators are often hindered by the weak, unspecific interactions between molecules and insulating surfaces, which leads to surface diffusion and assembly of disordered aggregates. Furthermore, insulating substrates require characterization approaches, which are rather independent of the conductivity. One way is the usage of ultra-thin insulating films which was shown to allow even orbital imaging by low-current STM [7]. An alternative way is the usage of non-contact atomic force microscopy (nc-AFM) [8], which cannot only image molecules on insulating films or bulk insulator surfaces but is also able to determine mechanical and electrical properties at the submolecular scale, e.g., [9–16]. Even vertical and horizontal manipulation processes as well as measurements of single molecules on insulating surfaces at room temperature are possible [17–21]. This process might also be important for molecular electronics since molecules have to be arranged or contacts have to be moved to the molecules. In a seminal work, Gross and co-workers [22] demonstrated that nc-AFM is not only able to image single atoms and molecules, but is also able to detect the charge state of adatoms on thin insulating films. The authors used bias spectroscopy measurements related to Kelvin probe force microscopy (KPFM) [23–25] and determined the so-called local contact potential difference (LCPD) [26].

The achievements in scanning probe microscopy have tremendous consequences for the field of molecular electronics [27]. Molecules can now be imaged down to the submolecular level; even on insulating substrates, single metal atoms can be manipulated to certain positions and act as electrical contact to the molecule; and finally, the electrical and mechanical properties can be determined also under external excitation. Figure 1 summarizes and illustrates these advances. As shown in the figure, the AFM tip or an external excitation can alter the charge state of the

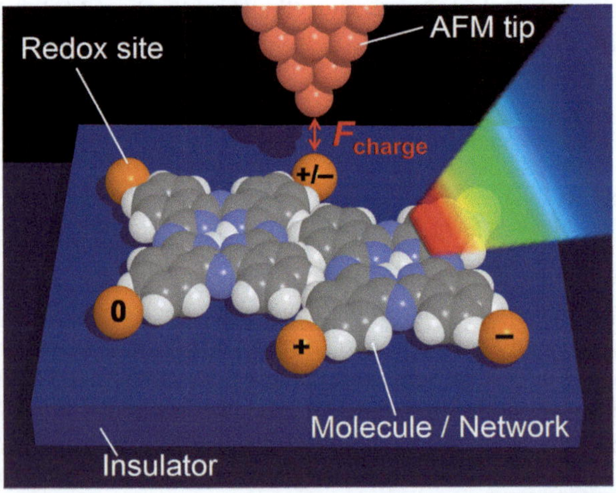


**Fig. 1** A single molecule or molecular network on an insulator can be probed by the attachment of metallic adatoms, where single electron charges can be excited by incident photons. Subsequently, the charge transport can be investigated by the measurement of charge states of all the terminal adatoms or directly within the molecule by scanning probe microscopy. Image adopted from [27]

molecule which can afterward be measured by the change of the charge states of the others terminals. This leads to a direct insight into the energy adsorption and propagation of charges in the molecule or a molecular network. Such a change of the charge state of a molecule is in principle an artificial way of a local chemical reduction or oxidation process. Just recently, local Raman scattering measurements at submolecular scale have been demonstrated [28].

In this chapter, the results in the field of advanced atomic force microscopy for high-resolution detection and manipulation of single molecules done at the University of Basel in the Nanolino-group [29] are summarized and discussed. Started some years ago, porphyrin-type molecules were functionalized to allow the stable adsorption of molecular assemblies on alkali halide surfaces (Sect. 2). The formation, manipulation and even the contacting of such wires by non-contact atomic force microscopy (nc-AFM) at room temperature were studied [11, 24, 30–33]. Finally, even a single truxene molecule could be stabilized at room temperature on a KBr surface at a step edge [34, 35]. Section 3 summarizes results acquired by low-temperature combined nc-AFM and STM measurements. At low temperatures, the mobility of the molecules can be reduced so that local elasticity, manipulation, and finally also the determination of electrostatic properties (Sect. 4) are possible [36–43].

## 2 Porphyrin-Based Molecular Assemblies

$\pi$-conjugated molecules such as porphyrins with tunable electronic properties are appealing building blocks for the construction of functional materials with exceptional electrochemical and photophysical properties [44]. All porphyrins absorb light in the visible range and therefore appear colored and are interesting candidates for photovoltaic applications [45]. The chemical structure of a porphyrin molecule is based on a macrocycle having four pyrrole rings cyclic linked by four methyne groups to build an extended aromatic $\pi$-system with 18 $\pi$-electrons per core. The core has a fourfold symmetry and can be augmented with functional groups in many ways. Furthermore, the two hydrogen atoms in the middle can be substituted by a metal atom (e.g., Fe, Mg, and Zn, …) having profound influence on the electrochemical activity. This might also explain why porphyrin molecules are very crucial building blocks in nature, e.g., the most important ones are involved in photosynthesis and the transport of oxygen. In the metabolism of mammals, oxygen is transported by red blood cells, the hemoglobin. The major part of it is the HEME group, an iron-(II)-substituted porphyrin. The photosynthesis taking place in plants uses chlorophylls which are magnesium-(II)-chlorin complexes with chlorin being a porphyrin derivate. To allow photosynthesis, the porphyrins absorb red light which is responsible for the green color of all plants [44].

The growth of ordered molecular structures on insulators was in the beginning limited due to a lack of well-adapted molecules. By developing specific cyanophenyl end groups and link them to the porphyrin core [46], these could be solved. In the following, structural, mechanical, and electrical measurements of well-ordered

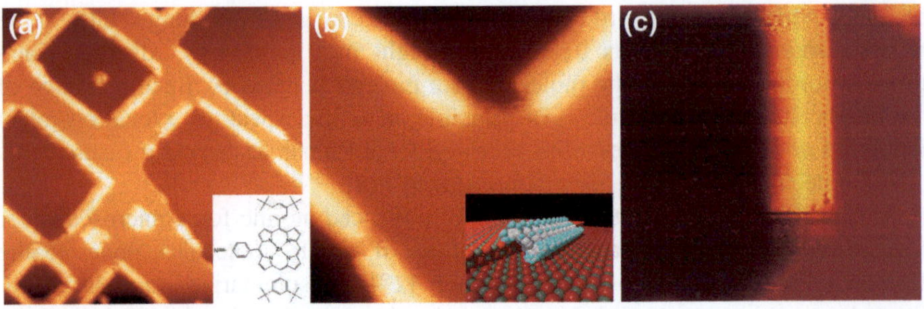


**Fig. 2** **a** nc-AFM measurement of single atomic KBr step edges decorated with cyanoporphyrin molecules (*inset*), size = 100 × 100 nm$^2$. **b** Wire-like molecular structures which are stabilized by an intermolecular interaction (π–π stacking) and directed electrostatic interaction with the alkali halide substrate. **c** A high-resolution nc-AFM measurement (size = 22 × 22 nm$^2$) shows single-molecular resolution, while some molecules are already removed from the wire [30]

nanostructures of meso-(4-cyanophenyl)-substituted Zn(II) porphyrin molecules (inset of Fig. 2a) will be described. As substrates, mainly alkali halide single crystals are used, which offer an easy-to-prepare, insulating, and atomically flat surface with well-known properties.

The porphyrin assemblies are formed along step edges and along specific directions of KBr(100) and NaCl(100). Short molecular wires, ringlike structures, long molecular wires (>250 nm), and oriented multiwires are observed by high-resolution nc-AFM at room temperature. The measured intermolecular distances of 0.5–0.6 are similar to natural and biomimetic light-harvesting structures and indicate π–π stacking of the porphyrin rings. The immobilization of the porphyrin molecules is due to electrostatic interaction between the cyanophenyl group and the ionic surface, which yields stable, defect-free molecular aggregates [30, 31]. A similar behavior was also found for porphyrins deposited on thin alkali halide films [47]. In contrast to the parallel orientation of porphyrin molecules deposited on metal surfaces [48], they are tilted on KBr by 35°–45° with respect to the surface. The electrical decoupling as well as the stacking of these porphyrine wires are necessary requirements for the application as light-harvesting structures. Such structures are based on the formation of excitons along the wires and might be used in molecular electronics or machinery. Furthermore, contacting by metal clusters [11, 24, 33], controlled cutting and self-healing processes [32] have been observed and will be summarized.

## *2.1 Deposition of Porphyrins on Insulating Surfaces*

In 2004, Nony et al. [9] showed the confinement of chloro-[subphthalocyaninato] boron-(III) (SubPc) molecules in nanometer-sized pits on KBr(100) which were related to the interaction of the molecular dipole (5.4D) with the electrostatic corrugation of the surface. To enhance the molecule–molecule interaction while still providing a large molecular dipole moment, a special cyanoporphyrin molecule was synthesized and the chemical structure is shown in the inset of Fig. 2a. The core of the molecule is apolar, while the strong negative polarization of the nitrogen atom induces a dipole moment of 0.36 D allowing to interact with the alternating charges of the ionic surfaces. Porphyrin molecules tend to π–π stack allowing a reasonably strong intermolecular interaction for the stabilization of self-assemblies [30].

Figure 2a shows an nc-AFM image of an electron-irradiated KBr(100) surface [49, 50] at a constant frequency shift of $\Delta f = -20$ Hz, after the evaporation of cyanoporphyrin molecules. The structure of the irradiated pattern is still apparent, and decorated steps are clearly visible. Atomic resolution was obtained on flat terraces, inside the pits, and sometimes also close to the molecular structures [11, 31, 47]. This large-scale topographical image highlights also that only the straight step edges are decorated by porphyrins. As soon as the steps are rounded, the molecular wire becomes interrupted. In Fig. 2b, c, more details about the molecular wire itself are shown. In the lower part of Fig. 2c, an instability is observed where part of the molecular wire is removed, revealing the KBr step underneath it. A model of the molecular arrangement at the ionic step edges reveals the strong intermolecular stacking combined with the electrostatic interaction between the cyanophenyl groups and the alternating ionic potential at the step edges. The two 3,5-di(tert-butyl)phenyl side groups of the molecule are determining the tilting of the wire parallel and perpendicular to the step edge. Along one and two monolayer step edges, single-molecular wires are found, while at higher steps, disordered aggregates of molecules appear [47].

For larger step edges, disordered agglomerates of porphyrin molecules have been observed. Figure 3a shows such agglomerates on a stepped KBr(100) surface which acting as nucleation sites for ordered structures of parallel stacked wires. These multiwires were found to be mainly oriented along the [110]-direction of the KBr substrate. The number of parallel stacked wires can be controlled by the deposition rate. Assemblies with sizes from a few parallel wires up to several hundred nanometer wide structures are possible [31]. Figure 3b shows a high-resolution nc-AFM image of such parallel molecular wires separated by $2.4 \pm 0.2$ nm. This separation corresponds to approximately five lattice spacings of the substrate. In a work of Pšenčík et al. [51], distances between different bacteriochlorophyll stacks in natural chromosomes of 2.1–3.0 nm have been determined which is in the same order of magnitude as observed by Maier et al. for the porphyrin assemblies. However, unlike at the step edges in [100]-direction, the intermolecular distance ($0.56 \pm 0.01$) within a wire differs from the KBr lattice spacing of 0.47 between two $K^+$ ions along the [110]-direction. This indicates that the dimensions of the molecule require a larger separation than the $K^+$ intervals provide. However, the partially negative-charged cyanophenyl groups are forced into position by the alternating $K^+$ and $Br^-$ ions along the step edges in [100]-direction.

In [110]-direction, a positive potential well is formed by the potassium ions which are not interrupted by bromide ions. This allows the cyanoporphyrins to adjust to their thermodynamically preferred intermolecular spacing while stabilizing

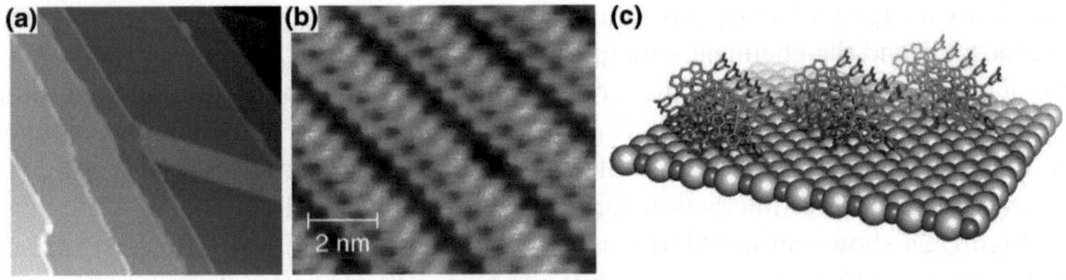


**Fig. 3** **a** nc-AFM topographical overview of atomic KBr step edges decorated with porphyrin molecules and a molecular carped (multiwire) in [110] substrate direction. **b** Intramolecular resolution of such a wire and parallel rows of the stacked molecules are visible. **c** Schematic model of the molecular multiwire on KBr [31]

in the positive potential well along the $K^+$ ions by the partial negative charge of the cyano end groups. Figure 3c shows the proposed multiwire arrangement in [110]-direction taking into account the sterical hindrance of the side groups and the measured intermolecular distances. Compared to the wires at the step edges, the angle between the porphyrin cores and the wire axis of the multiwires is enlarged to (43°–45°). It was concluded that the multiwires are attached by the cyanophenyl groups pointing downwards to the surface and the bulky side groups standing out on one side as depicted by the asymmetric shape of the molecule in the high-resolution image (Fig. 3b). Depending on the tilt angle of the stacks, monolayer heights between 1.5 and 2.0 nm were measured.

We showed further that the direction of the wires can be controlled by the atomic structure of the substrate so that complex molecular assemblies can be formed. NaCl is chemically and physically similar to KBr and is therefore an interesting sample to investigate the influence of the lattice distance of the substrate to the molecular self-assemblies. The lattice constant of NaCl is 5.65 Å which is very close to the spacing of 5.6 Å between the stacked molecules. An overview nc-AFM topography image of the molecular assemblies on NaCl(100) is shown in Fig. 4a. The step edges show no ordered molecular decorations and have no specific direction; however, single-wire decoration as on the KBr(100) has been found along steps in [100]-direction. Furthermore, a large amount of multiwires growing across the terraces from kink sites at step edges was observed. A major difference between the self-assemblies on KBr and NaCl is the formation of crossing molecular carpets or networks of molecular wires which is shown in Fig. 4a. The observed angles between the wires are 90° and 45° indicating a growth-oriented in all major crystallographic surface directions. The shape and thickness of the molecular structures do not differ along the ⟨100⟩ and ⟨110⟩ directions. The intermolecular distance of 5.65 Å making the wires along the ⟨100⟩ directions on NaCl more favorable compared to KBr and the molecules is most certainly adsorbed at every sodium atom. Furthermore, wires along the ⟨110⟩ directions can still grow from kink sites or wire junctions, which is depicted in Fig. 4b, c for the structural models of KBr(001) and for NaCl(001).

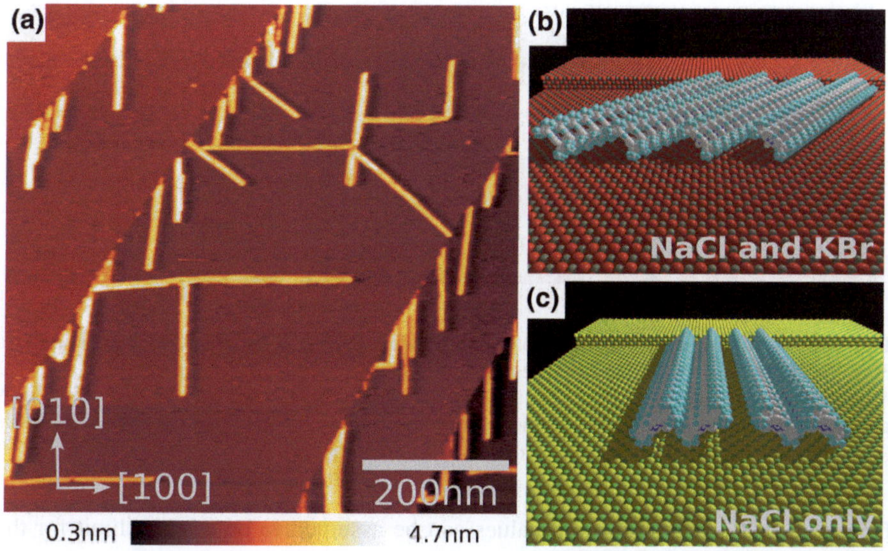


**Fig. 4** **a** nc-AFM topography of the cyanoporphyrin molecular wires on NaCl. The molecular wires are oriented along the [100]- and [110]-directions. Schematic views are depicted in **b** and **c** [47]

## *2.2 Contacting Porphyrin Structures*

The previous section showed the possibility to form and control the growth direction of molecular wires along step edges and across terraces. A requirement in molecular electronics is the contacting of such molecular structures with metal contacts. Evaporated gold forms small clusters on KBr and NaCl with a visible diameter of 2–10 nm and a height of roughly 1.5 nm. Figure 5a shows a measurement of such Au clusters and molecular assemblies of porphyrin wires [24]. The zoomed area in (b) highlights one wire contacted at both ends by a gold cluster. The LCPD [26] image in Fig. 5c resolves differences among the KBr surface, the Au nanoclusters, and the molecular wire. Between the KBr surface and the Au nanoclusters, differences of ≈250 mV and between Au nanocluster and the molecular wire of ≈220 mV were measured. The local dipole moment of the molecules defines the LCPD values at the wire which was measured to be $\Delta V_{\mathrm{LCPD}} = 180 \pm 20$ mV.

The LCPD can be related to the difference in dipole moment densities ($\Delta p$) by $\Delta V_{LCPD} = 1/\varepsilon_0 \Delta p$ which results in 0.5 D nm$^2$. Taking a molecule density of 0.72 nm$^{-2}$ [31] results in an average dipole moment density difference per molecule of 0.36 D. This is a reasonable value in comparison with the absolute dipole moment of a single porphyrin of 4.37 D. However, the well-known averaging effect [52] in KPFM measurements at very close proximity to the surface (<1 nm) has to

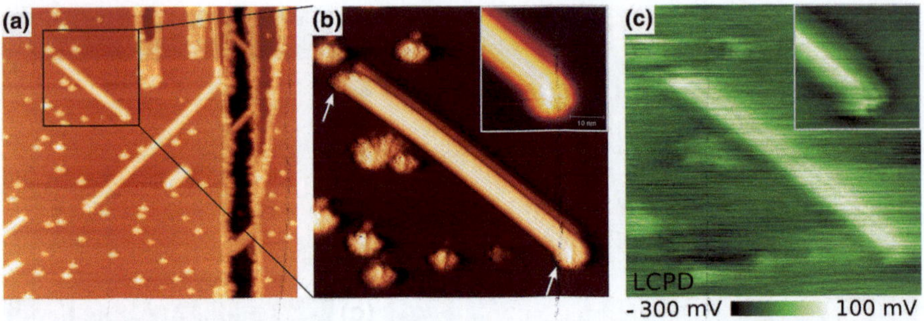


**Fig. 5** nc-AFM measurements of porphyrin multiwires on Au-decorated KBr(100). **a** shows a $300 \times 300$ nm$^2$ scan with various wires along the [110]-direction. **b** Topography and **c** LCPD measurements give a closer look of the region ($100 \times 100$ nm$^2$) marked by a *black frame* in **a**. A porphyrin multiwire assembly is formed between two Au clusters marked in **b**. The *insets* ($30 \times 30$ nm$^2$) highlight details of the lower end of the wire [24]

be considered and the absolute values can be assumed to be a lower limit for the dipole moment.

Even so the molecular structures can be contacted by metal clusters, an electrical readout is still not possible. The goal of the next study was therefore the electrical contacting of such molecular porphyrin wires. For that purpose, electrodes using nanostencil lithography (NL) were fabricated, which circumvents the use of resist layers and etching steps [53]. Building up the molecular wires and metal electrodes on a crystalline insulating layer required a complex sample fabrication, involving several evaporation and annealing steps. The starting point was an n-doped (001)-oriented GaAs wafer for subsequent use as a conductive substrate for the electrodes. A NaCl layer having a thickness of approximately 2.5 nm (sufficient for efficiently suppressing electron tunneling) was evaporated through a stencil mask. The porphyrin molecules and some Au cluster were evaporated onto the entire sample. Metal electrodes were fabricated by using the same mask as used for NaCl evaporation, but with a lateral offset. This produced a metal pad that partly overlapped with the NaCl pad. Electrical contact of the metal pad was established to the GaAs substrate in the region without overlap with the NaCl layer. As electrode material Cr was used, diffusion effects could be strongly minimized compared to Au electrodes. To facilitate the contacting of molecular wires, dendritic apertures were cut into the edge of the stencil membrane to form structured metal electrodes on the NaCl layer (cp. Fig. 6a) [33].

Figure 6a, b shows the topography and the contact potential difference (CPD) around an area near an electrode structure of such a sample, respectively. A measurement with increased resolution of a porphyrin multiwire contacted on one side by an Au cluster and on the other to the Cr electrode is shown in Fig. 6c. The wire was approximately 100 nm long with a height of 1 nm. This multiwire was built up from several individual molecular wires, as we deduced above, cp. Fig. 3. Taking the width of approximately 10 nm into account, the multiwire consisted of 4–5

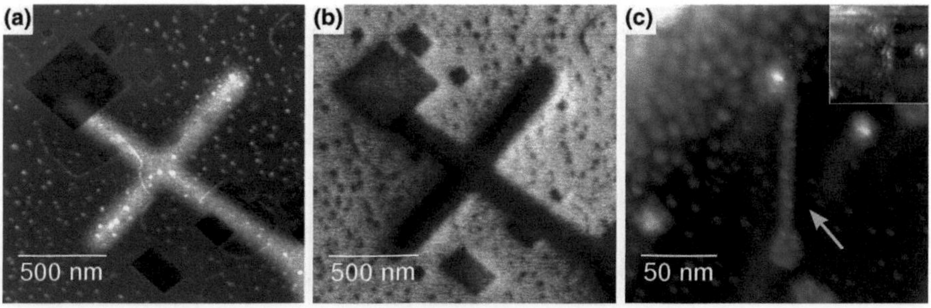


**Fig. 6** **a** Cr electrodes are evaporated after the molecular wires on top of a NaCl thin film on a GaAs substrate. The molecules were evaporated first, then the sample was annealed to 480 to induce the formation of molecular wires, and finally, Cr electrodes were evaporated. **b** shows the simultaneously acquired CPD image highlighting the different materials (ΔCPD = 720 mV). Molecular wires were also found near the electrodes as highlighted by an *arrow* in **c**. Bias spectroscopy was performed along the red dotted line, however the wire was destroyed afterward, see inset image [33]

single-molecular wires. To test the conductivity distance, spectroscopy measurements along the structure have been performed using the tip as the second electrode. At each measurement point (red dots along the wire in Fig. 6c), the tip was moved 23.5 nm toward the surface, measuring frequency shift, amplitude, dissipation, and current values. A sample bias voltage of $V_{\text{bias}} = 2.5$ V was applied. Above the Cr electrode, a current of up to 40 pA was measured, whereas above the molecular wire, no current could be detected. However, the molecular structure was completely damaged after the measurement (inset of Fig. 6c) most probably due to entering the repulsive regime of the interaction forces during the spectroscopic measurements. The measured onset of the current above the Cr electrode indicates that the electrode was connected successfully via the GaAs sample and also that such conductance measurements are in principle feasible.

## 2.3 Cutting and Regeneration of Porphyrin Structures

Already in the first measurement of porphyrin wires, it was observed that the structures can be easily destroyed by the AFM tip (cp. Fig. 2c). Figure 7 shows a series of topographical measurements of an intentionally deformed porphyrin multiwire obtained by nc-AFM at room temperature with an amplitude of 5.0 nm at a frequency shift of −5.1 Hz. In order to avoid accidental tip-induced modifications, the surface was scanned at a relatively large tip–sample distance ($\gamma = -0.5\ \text{fN}\sqrt{\text{m}}$). After Fig. 7a was obtained, the wire was cut by deactivating the tip–sample distance controller eleven times at the indicated positions. The multiwire was divided into three parts but surprisingly, in the next measurement, shown in Fig. 7c, the wire grew, and the gap between the middle and upper parts

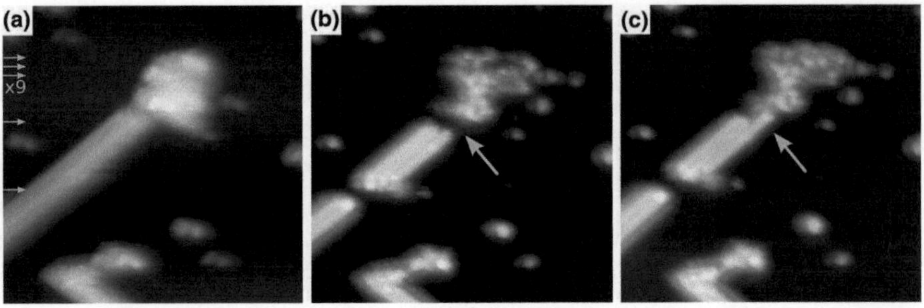


**Fig. 7** Intentional deformations of a porphyrin multiwire connected to an Au cluster ($90 \times 90$ nm$^2$). **a** Before cutting with the AFM tip, **b** with two distinct cuts, and **c** self-healing at room temperature [32]

became smaller, indicated with an arrow. Since no significant instability of the cantilever oscillation was observed, the regeneration process seems not be related to tip-induced modifications (i.e., deposition of molecules from the tip covered by molecules). It is most likely that mobile molecules, existing on the KBr(100) surface, self-assembled to close the gap in the wire. This behavior might be very important to achieve highly robust and self-healing future molecular devices.

## 3 Mechanical Properties

A general strategy allowing immobilization of single molecules on an insulating substrate at room temperature is a prerequisite for molecular electronics. In Sect. 2 the self-assembly and the stabilization of porphyrin molecules was presented. The stabilization at room temperature occurred due to $\pi$–$\pi$ interaction between the cores of the molecules, while the substrate interaction controlled the absorption of the assemblies on ionic surfaces [47]. To allow an even stronger molecule surface interaction, stronger localized and directed dipole moments are necessary. Custom-designed syn-5,10,15-tris(4-cyanophenylmethyl)truxenes (inset of Fig. 8a) contain three flexible cyanobenzyl groups, possessing a large dipole moment, which immobilizes even single molecules at specific sites of the ionic substrate allowing them to be imaged by nc-AFM at room temperature [34].

Figure 8b shows molecular dynamics (MD) calculations of the truxene molecule attached to a kink of a KBr step edge. The combination of the three flexible binding groups interacting independently with the surface is crucial for anchoring and enhances the binding energy at the kinks, while the molecules at the bare step edges are at room temperature still mobile, leading to the fuzzy contrast in Fig. 8a. The molecule–substrate interaction was even found to be large enough to restructure the ionic surface providing anchoring sites for the single molecules. Figure 8c shows a former rectangular KBr pit which was reconstructed while evaporating the truxene

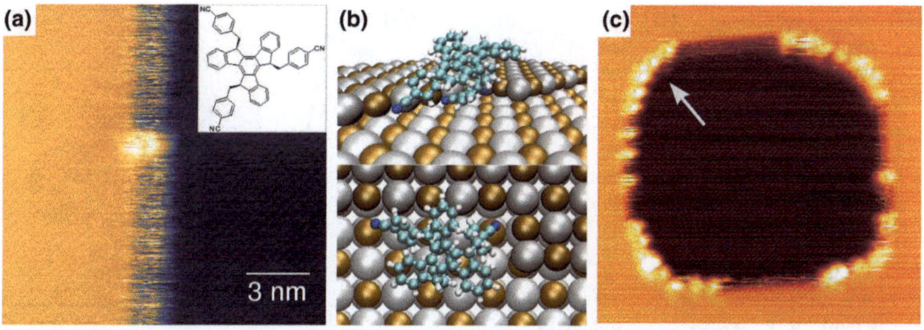


**Fig. 8** **a** Single truxene molecule (*inset*) stabilized at a kink of a KBr step edge. The fuzzy contrast at the atomic step edge is induced by mobile molecules [34]. **b** MD calculations confirm the attachment of the cyanobenzol groups to the kinks and the overall stabilization of the molecule. **c** During deposition of the molecules, the surface of KBr is adopted to provide sufficient adsorption sites for the molecules. Former rectangular pits are transformed to round-shaped structures [35]

molecules to the surface [35]. So the mechanical properties such as elasticity and mobility of molecules on the surfaces are crucial physical parameters to understand, control and predict molecular self-assemblies for applications.

## 3.1 Elasticity of Single Molecules

Recently, atomic force microscopy has made remarkable progress in imaging organic molecules allowing inner structure observations within molecules [15]. Although the atoms and bonds of flat molecules have been resolved, the study of three-dimensional molecules such as $C_{60}$ and the measurement of intrinsic mechanical properties at the atomic level within a molecule is still challenging. We presented submolecular resolution of the $C_{60}$ molecule obtained by combined nc-AFM and STM in 2011 revealing its chemical structure, describing unambiguously the $C_{60}$ orientations on the substrate as well as local elasticity measurements. The same molecule was used by Gross and co-workers [54] to discriminate the bond order of individual carbon–carbon bonds in polycyclic aromatic hydrocarbons and fullerenes. Moreover, three-dimensional spectroscopic measurements of the interaction force and the tunneling current above single molecules allowed us to obtain a detailed insight in the observed contrast. By our technique, we could show site-specific stiffnesses within the molecular structure which are related to the local elastic properties of the fullerene structure and confirms its predicted robustness at the atomic scale which is not possible by STM alone. Recently, the molecular stiffness of a single CO molecule was determined with a similar setup [55].

Systematic three-dimensional (3D) spectroscopic nc-AFM measurements at 77 on $C_{60}$ molecules being adsorbed on top of self-assembled $C_{60}$ islands have been

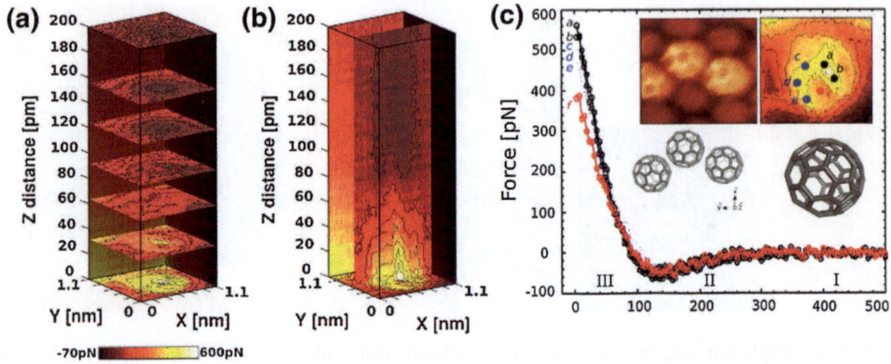


**Fig. 9** **a** 3D force field $F_z(x, y)$ acquired by nc-AFM at different tip–sample separations and **b** $F_z(x, z)$ cross section taken at the center of the molecule, revealing the contrast evolution with respect to the tip–molecule distance $z$. Parameters, $1.1 \times 1.1 \times 0.5$ nm$^3$ with $60 \times 60 \times 128$ data points, $V_{tip} = 300$ µV, $A = 60$ pm [37, 40]

performed [37]. Figure 9a, b illustrate the evolution of the 3D force field above a single molecule $F_z(x, y)$ at different tip–sample distances as well as a vertical cross sections $F_z(x, z)$ taken at the center of the molecule, respectively. For tip–sample distances smaller than 200 pm, the detected contrast appears as a homogeneous dark protrusion above the $C_{60}$ and is induced by attractive forces of up to −70 pN. Simultaneously, small tunneling currents of 10–20 pA have been detected. For tip–sample distances smaller than 100, the observed contrast in the $F_z(x, y)$ field shows a local increase up to positive values with submolecular resolution. This shows that a variation of the tip–sample distance of only 100 pm is sufficient to induce the disappearance of the inner molecular structure during imaging. The inset of Fig. 9c presents the constant-height force map of the closest tip–sample distance, which reveals the chemical structure of the $C_{60}$ with an edge between polygons on top of the molecule. The deduced orientation of the $C_{60}$ is just plotted below [37, 40].

The analysis of individual force curves is shown in Fig. 9c. Specific sites (see right inset of Fig. 9c) corresponding to carbon sites (marked a–e) and the center of a carbon ring (f) were analyzed. In region I and II, more far away from the surface, the curves are nearly indistinguishable as expected for long-range attractive forces. For tip–sample distances smaller than 100 pm (region III), differences induced by the repulsive force regime have been observed. While the total interaction force obtained in positions a and b reaches a maximal value of 550–570 pN, the value is slightly lowered to about 460–500 pN due to a slightly larger tip–sample distance for similar carbon sites c, d, and e. Finally, the center of the carbon ring (site f) exhibits the smallest repulsive force of 390 pN.

The slopes of the force–distance curves in the repulsive force regime can be interpreted as the local elastic properties appearing between the structure of the $C_{60}$ molecule and the Cu-terminated tip. Above carbon atoms (sites a–e), this vertical

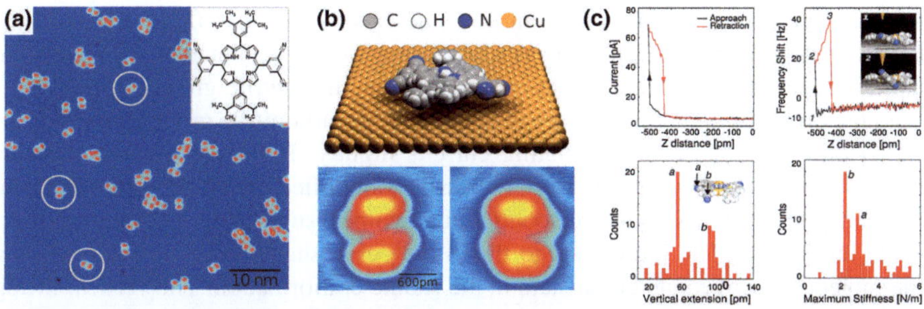


**Fig. 10** **a** STM topography in constant current mode of H2TBCPP on Cu(111) showing individually adsorbed porphyrin molecules oriented along three directions of the substrate (*circles*). **b** Schematic view of the saddle conformation and STM images of the two enantiomers [56]. **c** Individual approach–retract curves taken above a single nitrogen atom and the determined vertical extension as well as the tip–sample stiffness during the switching process [39]

force gradient is found to be 7–9 N/m, whereas above the center of the carbon ring (site f), this value is lowered to 4 N/m. Since only a minor dissipation contrast was detected, reveling a pure elastic interaction as well as a stiffer tip apex was assumed, the observed site-dependent force variations have been related to submolecular stiffness variations of the $C_{60}$ structure.

Another example of the local deformation of a single molecule on a surface is the study presented by Pawlak et al. [39]. $H_2$TBCPP porphyrin molecules (chemical structure in inset of Fig. 10a) were thermally evaporated on a cold Cu(111) substrate ($T$ = −190 K. Figure 10a shows individually adsorbed molecules oriented along the [110]-direction or an equivalent symmetry direction (marked by circles). The di-tert-butylphenyl groups of each molecule appear as bright protrusion, while the di-cyanophenyl groups are darker which is described also for similar molecules as saddle conformation (cp. Fig. 10b upper image) [57]. The two di-cyanophenyl groups are slightly rotated compared to the molecular center leaving one carbonitrile group closer to the Cu(111) substrate than the other and resulting in two enantiomers as shown in the lower part of Fig. 10b.

The submolecular elastic properties of this molecule have been studied again by combined 3D force and tunneling spectroscopy at low temperatures ($T$ = 4.8 K). The frequency shift data set revealed four distinct submolecular regions at the periphery of the molecule with a strong positive frequency shift, the rest of the values is negative hence still in the attractive force regime. These points can be related to the carbonitrile groups of the porphyrin molecule. The forward and backward data sets revealed a strong hysteresis in these areas (transition 1–2). The same phenomenon is visible in the tunneling current channel, showing an abrupt jump to higher currents. The corresponding distance-dependent $\Delta f$ and $I_t$ curves are shown in the upper panel of Fig. 10c. Both curves clearly indicate a hysteresis loop at the point of closest tip–sample distance between the forward and backward scan, while further away they match perfectly. Therefore, tip changes can be excluded as well as a strong influence by repulsive forces.

The oscillation amplitude used in this studies was $A = 40–60$ pm and the hysteretic behavior was attributed to the formation of a junction between the Cu-terminated tip and the N atom of the targeted carbonitrile group. Upon retraction, the bond formation is strong enough to lift the carbonitrile group (transition 2–3). Within this region, the tunneling current slightly decreases, because the carbonitrile group is disconnected from the surface and lifted up. At the same time, the frequency shift increases in a discontinuous manner, suggesting a strong tip–molecule interaction. The jump back to the initial curve shape (point 3) indicates that the molecule recovers its initial state in the saddle conformation. The vertical lifting process happens independently of the targeted carbonitrile group and is fully reversible.

To extract the local stiffness $k_{ts}$ of this submolecular contact, the length of the hysteresis loop ($z$-distance between point 2 and 3) as well as the maximum frequency shift were determined in a statistical analysis. The vertical extension as well as the stiffness is plotted in the lower diagrams of Fig. 10c. Both histograms have two pronounced peaks (a and b) attributed to two different lifting processes, one for a carbonitrile group which initially have been more far away from or closer to the surface, respectively. The maximum stiffness determined for these two distinct contacts is $k_{ts(a)} = 2.9$ N/m and $k_{ts(b)} = 2.0$ N/m.

## 3.2 *Manipulation of Single Molecules*

Directed molecular manipulation is a key step toward the understanding of molecular electronics as well as to build up single-molecular devices. To initiate and control molecula motion on a surface mechanical, chmical, or electrical interactions are used. However, as was described in Sect. 3.1, elasticity of molecular structures plays a major role in all diffusion processes. In response to the just above discussed local tip–sample force, the molecular structure is elastically deformed and leads finally also to a controlled rotation [37, 39].

If the tip is approached slightly further (20 are sufficient!) compared to the separations that induce only an elastic deformation, the bond between the carbonitrile group and the Cu tip apex is sufficiently strong to induce a rotation of the molecule by 60°. However, the exact lateral position of the tip is important for the strength of the interaction and the bond might already not be strong enough anymore to induce the rotation. The target area for a clockwise rotation is highlighted by the red area in Fig. 11a. The rotation direction depends on the targeted carbonitrile group, for a clockwise rotation the red areas, while for an anticlockwise direction the green areas have to be approached by the tip. An example of both rotation directions is shown in the overview images in Fig. 11b. For all adsorption directions and enantiomers, this behavior was found to be valid success rate of 75 % [39].

Some details about this manipulation process can be gained by analyzing the respective distance-dependent measurements as presented in Fig. 11c. The tunneling current plotted in the upper diagram shows again the discussed

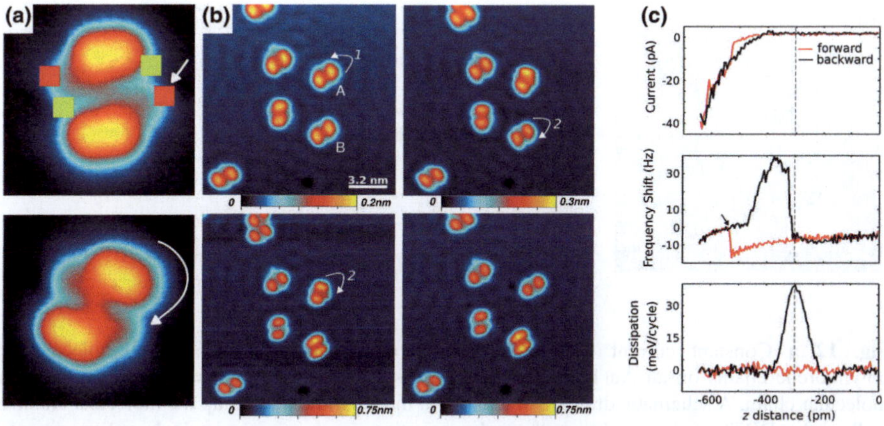


**Fig. 11** **a** STM images showing a single rotation of a H2TBCPP molecule on Cu(111), the *white arrow* shows the used position of the *z* spectroscopic curve for the clockwise rotation (*red areas*). Counterclockwise rotations can be induced at *green areas*. **b** Images of successive manipulation processes. **c** Typical distance-dependent curves of the tunneling current, the frequency shift, and the dissipation leading to a molecular rotation [39]

abrupt jump when contacting tip and molecule. However, the hysteresis is strongly distorted due to the rotation. For the frequency shift and the dissipation, significant features are visible in the retraction curves. The hysteresis in the frequency shift channel is extended, while the dissipation rises up to values of ≈40 meV/cycle. This peak in the backward dissipation signal is one of the main characteristics of a successful and force-induced rotation process. Similar to the elastic deformation, a junction is formed between the tip apex and the molecule upon approach. During retraction, first the di-cyanophenyl group is lifted up; however, the bond is stronger and the group is lifted up further. Upon this process, the second carbonitrile group is partly detached from the surface and laterally moved to a neighboring equivalent adsorption site. This detachment gives rise to stochastic changes of the tip–sample interaction and therefore to an increase of the frequency shift and the dissipation. Upon retraction of the tip and breaking of the tip–carbonitrile bond, the molecule relaxes to the new equilibrium position.

Another important manipulation process is a lateral manipulation of molecular structures for understanding their mechanical properties. The most fundamental and clean approach to understand these properties is a single-molecular wire pulled off a surface measured with atomic-scale resolution. Just recently, Kawai et al. [21] presented a study of the mechanical behavior of in situ polymerized fluorene wires with a length of 10–100 by pulling individual wires with a tip of a low-temperature nc-AFM. Figure 12a shows an Au(111) surface covered by molecular wires deposited by polymerization of a dibromoterfluorene (DBTF) precursor, schematically presented in the lower part of Fig. 12b [58]. The chemical structure and a high-resolution image of one end of the wire is shown in the inset. Most of the wires are physisorbed along the fcc stripes of the Au(111) herringbone structure at 4.8.

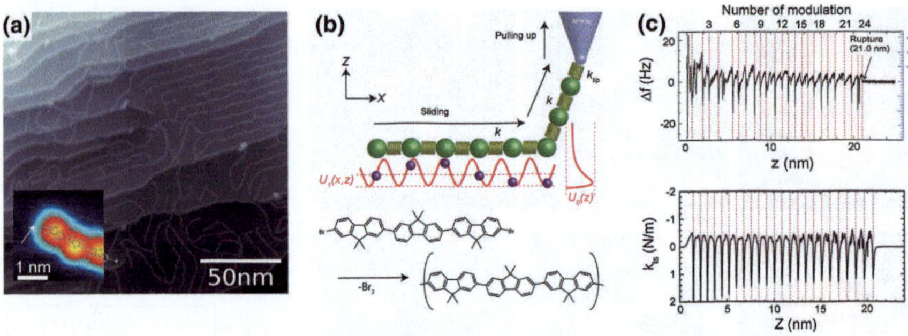


**Fig. 12 a** Constant current STM measurement of the polymerized DBTF molecules to polyfluorene chains on an Au(111). The *inset* shows a zoom-in STM image near one end of a molecular chain. A schematic drawing of a physical model while pulling up the molecular chain as well as the DBTF molecule during the polymerization process is shown in **b**. **c** Experimental frequency shift data measured while retracting the tip after contacting and pulling a chain and calculated force gradient [21]

By contacting the AFM tip to one end of the wire and pulling it up, the frequency shift at the resonance of the tuning fork sensor was recorded (Fig. 12c). The measured periodic variations of the force gradients correspond to the step-by-step detachment of fluorene units as shown by the schematic model in Fig. 12b. Calculations based on an extended Frenkel–Kontorova model reproduced the measured mechanical behavior and revealed that the stick forces and energy barriers are almost independent of the unit being detached.

All the presented examples demonstrate that the mechanical behavior of single molecules as well as low-dimensional molecular assemblies on surfaces can be characterized with atomic-scale precision in a control manner by high-resolution nc-AFM measurements.

# 4 Electrical Properties of Single Molecules

Also the electronic properties of single-molecular adsorbates play a decisive role in many fundamental processes as for instance in catalysis [59, 60], organic photovoltaics [61], and particularly also in molecular electronics. The adaptation of the energy alignment and local charging processes significantly modify the electronic structure of both the molecule and substrate. Combined nc-AFM and STM offers not only to image the topographical structure down to the atomic scale and measure, and manipulate mechanical properties as shown in the sections before, but also to determine local electronic properties such as the local work function or charge density. One of the major advanced microscopy techniques allowing to determine such properties is called KPFM [62]. In this section, we will just briefly introduce

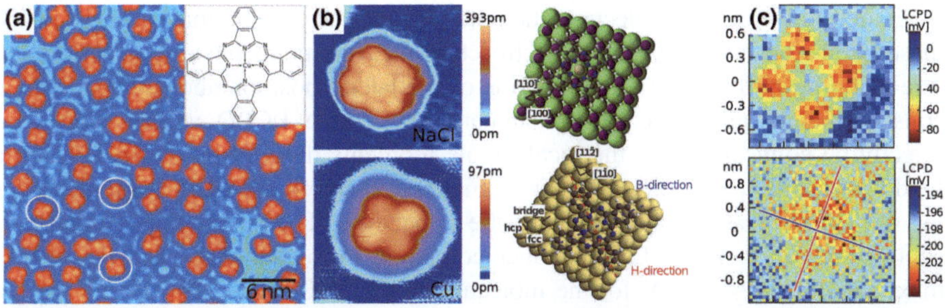


**Fig. 13 a** STM constant current image of CuPc molecules (*inset*) on Cu(111), three different orientations have been observed. **b** High-resolution constant current images and DFT-calculated adsorption geometries of CuPc on 2ML NaCl on Cu(111) (*top*) and bare Cu(111) (*bottom*). **c** LCPD showing intramolecular resolution of both configurations obtained by bias spectroscopy [63]

some recent studies focusing on submolecular resolution allowing to determine intramolecular charge densities.

Figure 13 shows the structure and local electronic properties of copper phthalocyanine (CuPc) molecules [inset of (a)] adsorbed on a metallic Cu(111) substrate, either directly, or on top of a decoupling NaCl bilayer [56, 63]. A comparison of these strongly and weakly interacting systems allows to directly conclude on structural distortions and charge distributions within the single molecules. While CuPc exhibits a fourfold $D_{4h}$ symmetry in the gas phase, it is reduced to twofold $D_{2h}$ symmetry by the adsorption on the metal surface, resulting in three different orientations on the Cu(111) substrate. For metal phthalocyanine molecules, it is assumed that an electron transfer from the metal surface to the molecule takes place [64, 65]. For CuPc on Cu(111), the overall charge transfer is assumed to be two electrons [66, 67]. However, in the presence of a NaCl bilayer, this charge transfer is affected and the electronic structure of the molecule remains nearly as in its gas phase [7].

In Fig. 13b, constant tunneling current images and DFT simulations of a single CuPc molecule adsorbed on a 2ML NaCl on Cu(111) (top) and bare Cu(111) (bottom) are shown. The fourfold symmetry for the molecule on the NaCl bilayer is as expected preserved indicating rather weak molecule–surface interactions. The DFT-calculated adsorption geometry in the lowest energy configuration reveals that in the energetically most favorable conformation, the central Cu atom is placed above a $Cl^-$ ion. Due to the coinciding fourfold symmetry of the NaCl layer and the molecule, all four benzopyrrole rings are located at equal adsorption sites. The reduced twofold symmetry, one pair of opposing benzo-pyrrole rings appears brighter than the other, of the CuPc molecule on the bare Cu(111) surface is clearly visible in the measurement below. The combination of the sixfold symmetry of the Cu(111) and the fourfold symmetries of the molecule resulting in a twofold symmetry of the combined system which is also visible in the DFT calculations.

To determine the local electrostatic potential and the distribution of charge inside these two setups, 3D local bias spectroscopy measurements have been performed. The mapping of the LCPD as a function of the lateral position above the surface at constant height allows to obtain subnanometer scale LCPD values which can directly be connected with the local dipole moments induced by the local electric field between tip and molecule/surface. Figure 13c shows the LCPD of both setups the one on the NaCl bilayer again in the upper row. The contrast above the molecule is shifted to lower LCPD values compared to the substrate, and the overall maximal shift is ≈ 80 mV for the molecule on the NaCl and only ≈10 mV on the bare Cu(111). However, the central Cu atom of the CuPc appears at more positive LCPD values than the benzopyrrole rings. The major difference of the CuPc adsorbed on Cu(111) beside the overall lower contrast is the completely broken symmetry in between the benzopyrrole rings. While the two rings along the red line feature similar contrast, one of the rings along the blue line features brighter contrast than the other.

A more negative LCPD value corresponds in the used setup to a more negative charge at the surface, here the benzopyrrole rings. For the CuPc on the NaCl layer, the charge distribution between the center Cu atom and the four benzopyrrole rings seems to be preserved, while the surface–molecule interaction in the case of the Cu (111) sample reduces this charge contrast by a different adsorption geometry but also by a charge transfer between them. The asymmetry within the LCPD contrast for CuPc on Cu(111) could also be proven by DFT calculations and is based on the fact that one benzopyrrole ring adsorbs above a fcc—while the other one above a hcp—hollow site. This gives rise to different molecule–substrate interactions at the two positions, and hence to different charge densities.

Another study was performed on a functionalized phenylethynylbenzene molecule (FFPB) providing an electron-donating and electron-accepting group (see inset in Fig. 14) [42]. The molecule was self-assembled on an Au(110)-(1 × 2)

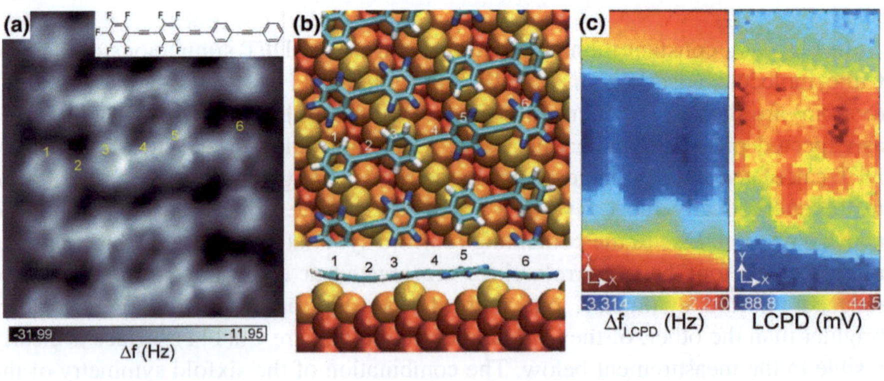


**Fig. 14** **a** Frequency shift image of a donor–acceptor molecule (FFPB, *inset* chemical structure) self-assembled on Au(110)-(1 × 2). **b** Side and top views of the relaxed molecule on the Au substrate using DFT. **c** A 3D bias voltage-dependent measurement revealed the frequency shift and the LCPD measured with an Au tip [42]

surface forming one-dimensional supramolecular structures. Figure 14a shows the complex chemical structure in the frequency shift image revealed by atomic force microscopy. It was observed that a major stability requirement of these self-assemblies is the formation of directional hydrogen bonds. A comparison of detailed geometrical deformations of the molecule on the substrate with density functional calculations is in perfect agreement with the measurements (cp. Fig. 14b). Furthermore, detailed bias spectroscopy measurements have been carried out providing the intramolecular charge distribution and allowed to identify the fluoro-substitution sites of each molecule. The data have been acquired by fitting the measured bias spectroscopy data to a parabola and calculating the LCPD and the corresponding frequency shift as shown in Fig. 14c. The LCPD contrast is governed by the local dipole and charge state, and the strongly electronegative fluorine atoms induce stronger local dipoles than the CH bonds. Therefore, the direction of each molecule could unambiguously be determined. However, the dipole of this molecule is located mainly parallel to the surface and the variations only appear due to the geometrical distortions and are influenced by the substrate interaction.

## 5 Summary

The presented results show that scanning probe microscopy techniques serve as unique investigation tools concerning fundamental questions related to molecules on surfaces. The deposition of molecules on insulating surfaces was demonstrated, and the in-detail characterization of the mechanical properties revealed a clear picture of the adsorption and stabilization process. Molecules with functionalized groups have been sensitized and allow together with an appropriate substrate the controlled growth of single molecules and various assemblies. The electrical characterization at the molecular scale is still in progress and a very active research field. However, the presented results already allow to determine intramolecular charge transfer processes and might also lead to the development of new functional molecules. Future studies include the simultaneous excitation by visible light to study also optically induced charge transfer processes within single molecules.

**Acknowledgments** The contributions from S. Maier, L. Zimmerli, S. Fremy, L. Gross, R. Pawlak, B. Such and S. Kawai are highly appreciated. Such research would not be possible without the interdisciplinary collaborations between chemists and physicists; therefore, many thanks to F. Diederich, A. Echavarren, S. Decurtins, Ch. Housecroft, and E. Constable as well as the involved students. Many thanks also to E. Meyer and A. Baratoff for the continuous support and stimulating discussions. The work reported in this chapter was financially supported by the Swiss Nanoscience Institute, the Swiss National Science Foundation, the Joint Swiss-Polish Research Programme PSPB-085/2010, and the Commission for Technology and Innovation CTI.

## References


1. Mirkin, C.A., Ratner, M.A.: Molecular electronics. Annu. Rev. Phys. Chem. **43**(1), 719–754 (1992)
2. Weibel, N., Grunder, S., Mayor, M.: Functional molecules in electronic circuits. Org. Biomol. Chem. **5**, 2343–2353 (2007)
3. Heath, J.R.: Molecular electronics. Annu. Rev. Mater. Res. **39**(1), 1–23 (2009)
4. Binnig, G., Rohrer, H., Gerber, C., Weibel, E.: Surface studies by scanning tunneling microscopy. Phys. Rev. Lett. **49**(1), 57 (1982)
5. Binnig, G., Rohrer, H., Gerber, C., Weibel, E.: Tunneling through a controllable vacuum gap. Appl. Phys. Lett. **40**(2), 178 (1982)
6. Binnig, G., Rohrer, H.: Scanning tunneling microscopy. Helv. Phys. Acta **55**(6), 726–735 (1982)
7. Repp, J., Meyer, G., Stojkovic, S., Gourdon, A., Joachim, C.: Molecules on insulating films: Scanning-tunneling microscopy imaging of individual molecular orbitals. Phys. Rev. Lett. **94**, 026803 (2005)
8. Binnig, G., Quate, C., Gerber, C.: Atomic force microscopy. Phys. Rev. Lett. **56**(9), 930 (1986)
9. Nony, L., Gnecco, E., Baratoff, A., Alkauskas, A., Bennewitz, R., Pfeiffer, O., Maier, S., Wetzel, A., Meyer, E., Gerber, C.: Observation of individual molecules trapped on a nanostructured insulator. Nano Lett. **4**(11), 2185–2189 (2004)
10. Fendrich, M., Kunstmann, T., Paulkowski, D., Moller, R.: Molecular resolution in dynamic force microscopy: topography and dissipation for weakly interacting systems. Nanotechnology **18**(8), 084004 (2007)
11. Glatzel, T., Zimmerli, L., Meyer, E.: Molecular assemblies on insulating ultrathin films analyzed by nc-AFM and KPFM. Isr. J. Chem. **48**(2), 107 (2008)
12. Milde, P., Zerweck, U., Eng, L.M., Abel, M., Giovanelli, L., Nony, L., Mossoyan, M., Porte, L., Loppacher, C.: Interface dipole formation of different ZnPcCl8 phases on Ag(111) observed by Kelvin probe force microscopy. Nanotechnology **19**(30), 305501 (2008)
13. Gross, L., Mohn, F., Moll, N., Liljeroth, P., Meyer, G.: The chemical structure of a molecule resolved by atomic force microscopy. Science **325**(5944), 1110 (2009)
14. Schütte, J., Bechstein, R., Rahe, P., Langhals, H., Rohlfing, M., Kühnle, A.: Single-molecule switching with non-contact atomic force microscopy. Nanotechnology **22**(24), 245701 (2011)
15. Gross, L.: Recent advances in submolecular resolution with scanning probe microscopy. Nat Chem **3**(4), 273–278 (2011)
16. Hauptmann, N., Mohn, F., Gross, L., Meyer, G., Frederiksen, T., Berndt, R.: Force and conductance during contact formation to a C 60 molecule. New J. Phys. **14**(7), 073032 (2012)
17. Morita, S., Yi, I., Sugimoto, Y., Oyabu, N., Nishi, R., Custance, O., Abe, M.: Mechanical distinction and manipulation of atoms based on noncontact atomic force microscopy. Appl. Surf. Sci. **241**(1–2), 2–8 (2005)
18. Sugimoto, Y., Pou, P., Custance, O., Jelinek, P., Abe, M., Perez, R., Morita, S.: Complex patterning by vertical interchange atom manipulation using atomic force microscopy. Science **322**(5900), 413–417 (2008)
19. Loske, F., Kühnle, A.: Manipulation of C60 islands on the rutile $TiO_2$(110) surface using noncontact atomic force microscopy. Appl. Phys. Lett. **95**(4), 043110 (2009)
20. Wagner, C., Fournier, N., Tautz, F.S., Temirov, R.: Measurement of the binding energies of the organic-metal perylene-teracarboxylic-dianhydride/Au(111) bonds by molecular manipulation using an atomic force microscope. Phys. Rev. Lett. **109**, 076102 (2012)
21. Kawai, S., Koch, M., Gnecco, E., Sadeghi, A., Pawlak, R., Glatzel, T., Schwarz, J., Goedecker, S., Hecht, S., Baratoff, A., Grill, L., Meyer, E.: Proceedings of the National Academy of Sciences (2014)
22. Gross, L., Mohn, F., Liljeroth, P., Repp, J., Giessibl, F.J., Meyer, G.: Measuring the charge state of an adatom with noncontact atomic force microscopy. Science **324**(5933), 1428–1431 (2009)

23. Kitamura, S., Suzuki, K., Iwatsuki, M., Mooney, C.B.: Atomic-scale variations in contact potential difference on Au/Si(111)-7x7 surface in ultrahigh vacuum. Appl. Surf. Sci. **157**, 222 (2000)
24. Glatzel, T., Zimmerli, L., Koch, S., Kawai, S., Meyer, E.: Molecular assemblies grown between metallic contacts on insulating surfaces. Appl. Phys. Lett. **94**(6), 063303 (2009)
25. Bocquet, F., Nony, L., Loppacher, C., Glatzel, T.: Analytical approach to the local contact potential difference on (001) ionic surfaces: Implications for Kelvin probe force microscopy. Phys. Rev. B **78**, 035410 (2008)
26. Enevoldsen, G., Glatzel, T., Christensen, M., Lauritsen, J., Besenbacher, F.: Atomic scale kelvin probe force microscopy studies of the surface potential variations on the $TiO_2(110)$ surface. Phys. Rev. Lett. **100**, 236104 (2008)
27. Meyer, E., Glatzel, T.: Novel probes for molecular electronics. Science **324**(5933), 1397–1398 (2009)
28. Zhang, R., Zhang, Y., Dong, Z.C., Jiang, S., Zhang, C., Chen, L.G., Zhang, L., Liao, Y., Aizpurua, J., Luo, Y., Yang, J.L., Hou, J.G.: Chemical mapping of a single molecule by plasmon-enhanced Raman scattering. Nature **498**(7452), 82 (2013)
29. Nanolino Group (2015). http://nanolino.unibas.ch/
30. Zimmerli, L., Maier, S., Glatzel, T., Gnecco, E., Pfeiffer, O., Diederich, F., Fendt, L., Meyer, E.: Formation of molecular wires on nanostructured KBr. J. Phys. Conf. Ser. **61**(1), 1357–1360 (2007)
31. Maier, S., Fendt, L.A., Zimmerli, L., Glatzel, T., Pfeiffer, O., Diederich, F., Meyer, E.: Nanoscale engineering of molecular porphyrin wires on insulating surfaces. Small **4**, 1115–1118 (2008)
32. Kawai, S., Maier, S., Glatzel, T., Koch, S., Such, B., Zimmerli, L., Fendt, L.A., Diederich, F., Meyer, E.: Cutting and self-healing molecular wires studied by dynamic force microscopy. Appl. Phys. Lett. **95**(10), 103109 (2009)
33. Gross, L., Schlittler, R.R., Meyer, G., Fendt, L.A., Diederich, F., Glatzel, T., Kawai, S., Koch, S., Meyer, E.: Contacting self-ordered molecular wires by nanostencil lithography. J. Vac. Sci. Technol. B **28**(3), C4D34–C4D39 (2010)
34. Such, B., Trevethan, T., Glatzel, T., Kawai, S., Zimmerli, L., Meyer, E., Shluger, A.L., Amijs, C.H.M., de Mendoza, P., Echavarren, A.M.: Functionalized truxenes: Adsorption and diffusion of single molecules on the KBr(001) surface. ACS Nano **4**(6), 3429–3439 (2010)
35. Trevethan, T., Such, B., Glatzel, T., Kawai, S., Shluger, A.L., Meyer, E., de Mendoza, P., Echavarren, A.M.: Organic molecules reconstruct nanostructures on ionic surfaces. Small **7**(9), 1264–1270 (2011)
36. Such, B., Glatzel, T., Kawai, S., Koch, S., Meyer, E.: Three-dimensional force spectroscopy of KBr(001) by tuning fork-based cryogenic noncontact atomic force microscopy. J. Vac. Sci. Technol. B **28**(3), C4B1–C4B5 (2010)
37. Pawlak, R., Kawai, S., Fremy, S., Glatzel, T., Meyer, E.: Atomic-scale mechanical properties of orientated C60 molecules revealed by noncontact atomic force microscopy. ACS Nano **5**(8), 6349–6354 (2011)
38. Fremy, S., Kawai, S., Pawlak, R., Glatzel, T., Baratoff, A., Meyer, E.: Three-dimensional dynamic force spectroscopy measurements on KBr(001): atomic deformations at small tip-sample separations. Nanotechnology **23**(5), 055401 (2012)
39. Pawlak, R., Fremy, S., Kawai, S., Glatzel, T., Fang, H., Fendt, L.A., Diederich, F., Meyer, E.: Directed rotations of single porphyrin molecules controlled by localized force spectroscopy. ACS Nano **6**(7), 6318–6324 (2012)
40. Pawlak, R., Kawai, S., Fremy, S., Glatzel, T., Meyer, E.: J. Phys.: High-resolution imaging of C 60 molecules using tuning-fork-based non-contact atomic force microscopy. Condens. Matter **24**(8), 084005 (2012)

41. Such, B., Glatzel, T., Kawai, S., Meyer, E., Turansky, R., Brndiar, J., Stich, I.: Interplay of the tip-sample junction stability and image contrast reversal on a Cu(111) surface revealed by the 3D force field. Nanotechnology **23**(4), 045705 (2012)
42. Kawai, S., Pina, C.M., Bubendorf, A., Fessler, G., Glatzel, T., Gnecco, E., Meyer, E.: Systematic study of the dolomite (104) surface by bimodal dynamic force microscopy in ultra-high vacuum. Nanotechnology **24**(5), 055702 (2013)
43. Pawlak, R., Glatzel, T., Pichot, V., Schmidlin, L., Kawai, S., Fremy, S., Spitzer, D., Meyer, E.: Local detection of nitrogen-vacancy centers in a nanodiamond monolayer. Nano Lett. **13**(12), 5803–5807 (2013)
44. Martin, R.E., Diederich, F.: Linear monodisperse π-conjugated oligomers: model compounds for polymers and more. Angew. Chem. Int. Ed. **38**(10), 1350–1377 (1999)
45. Yella, A., Lee, H.W., Tsao, H.N., Yi, C., Chandiran, A.K., Nazeeruddin, M., Diau, E.W.G., Yeh, C.Y., Zakeeruddin, S.M., Grätzel, M.: Porphyrin-sensitized solar cells with cobalt (II/III) based redox electrolyte exceed 12 percent efficiency. Science **334**(6056), 629–634 (2011)
46. Cheng, F., Zhang, S., Adronov, A., Echegoyen, L., Diederich, F.: Triply fused $Zn^{II}$-porphyrin oligomers: synthesis, properties, and supramolecular interactions with single-walled carbon nanotubes (SWNTs). Chem. Eur. J. **12**, 6062–6070 (2006)
47. Glatzel, T., Zimmerli, L., Kawai, S., Meyer, E., Fendt, L.A., Diederich, F.: Oriented growth of porphyrin-based molecular wires on ionic crystals analysed by nc-AFM. Beilstein J. Nanotechnol. **2**, 34–39 (2011)
48. Yokoyama, T., Yokoyama, S., Kamikado, T., Okuno, Y., Mashiko, S.: Selective assembly on a surface of supramolecular aggregates with controlled size and shape. Nature **413**, 619 (2001)
49. Such, B., Czuba, P., Piatkowski, P., Szymonski, M.: AFM studies of electron-stimulated desorption process of KBr(001) surface. Surf. Sci. **451**(1–3), 203–207 (2000)
50. Such, B., Kolodziej, J., Czuba, P., Piatkowski, P., Struski, P., Krok, F., Szymonski, M.: Surface topography dependent desorption of alkali halides. Phys. Rev. Lett. **85**(12), 2621–2624 (2000)
51. Pšenčík, J., Arellano, J.B., Ikonen, T.P., Borrego, C.M., Laurinmäki, P.A., Butcher, S.J., Serimaa, R.E., Tuma, R.: Internal structure of chlorosomes from brown-colored chlorobium species and the role of carotenoids in their assembly. Biophys. J. **91**(4), 1433–1440 (2006)
52. Glatzel, T., Sadewasser, S., Lux-Steiner, M.: Amplitude or frequency modulation-detection in Kelvin probe force microscopy. Appl. Surf. Sci. **210**(84), 84–89 (2003)
53. Zahl, P., Bammerlin, M., Meyer, G., Schlittler, R.R.: All-in-one static and dynamic nanostencil atomic force microscopy/scanning tunneling microscopy system. Rev. Sci. Instrum. **76**(2), 023707 (2005)
54. Gross, L., Mohn, F., Moll, N., Schuler, B., Criado, A., Guitian, E., Pena, D., Gourdon, A., Meyer, G.: Bond-order discrimination by atomic force microscopy. Science **337**(6100), 1326–1329 (2012)
55. Weymouth, A.J., Hofmann, T., Giessibl, F.J.: Quantifying molecular stiffness and interaction with lateral force microscopy. Science **343** (6175), 1120–1122 (2014)
56. Fremy, S.: The nature of condensed single molecules—local electronic and mechanical characteristics. Ph.D. thesis, University of Basel (2013)
57. Iancu, V., Deshpande, A., Hla, S.W.: Manipulating kondo temperature via single molecule switching. Nano Lett. **6**(4), 820–823 (2006)
58. Lafferentz, L., Ample, F., Yu, H., Hecht, S., Joachim, C., Grill, L.: Conductance of a single conjugated polymer as a continuous function of its length. Science **323**(5918), 1193–1197 (2009)
59. Sims, P.C., Moody, I.S., Choi, Y., Dong, C., Iftikhar, M., Corso, B.L., Gul, O.T., Collins, P. G., Weiss, G.A.: Electronic measurements of single-molecule catalysis by cAMP-dependent protein Kinase A. J. Am. Chem. Soc. **135**(21), 7861–7868 (2013)
60. Janssen, K.P.F., De Cremer, G., Neely, R.K., Kubarev, A.V., Van Loon, J., Martens, J.A., De Vos, D.E., Roeffaers, M.B.J., Hofkens, J.: Single molecule methods for the study of catalysis: from enzymes to heterogeneous catalysts. Chem. Soc. Rev. **43**, 990–1006 (2014)

61. Qu, J., Gao, B., Tian, H., Zhang, X., Wang, Y., Xie, Z., Wang, H., Geng, Y., Wang, F.: Donor-spacer-acceptor monodisperse conjugated co-oligomers for efficient single-molecule photovoltaic cells based on non-fullerene acceptors. J. Mater. Chem. A **2**, 3632–3640 (2014)
62. Glatzel, T.: Measuring Atomic-Scale Variations of the Electrostatic Force. Springer, Berlin (2011)
63. Fremy, S., Sadeghi, A., Baratoff, A., Goedecker, S., Glatzel, T.: Submitted for publication (2014)
64. Stadtmüller, B., Kröger, I., Reinert, F., Kumpf, C.: Submonolayer growth of CuPc on noble metal surfaces. Phys. Rev. B **83**, 085416 (2011)
65. Mugarza, A., Robles, R., Krull, C., Korytár, R., Lorente, N., Gambardella, P.: Electronic and magnetic properties of molecule-metal interfaces: transition-metal phthalocyanines adsorbed on Ag(100). Phys. Rev. B **85**, 155437 (2012)
66. Schaffert, J., Cottin, M.C., Sonntag, A., Karacuban, H., Bobisch, C.A., Lorente, N., Gauyacq, J.P., Möller, R.: Imaging the dynamics of individually adsorbed molecules. Nat. Mater. **12**(3), 223–227 (2013)
67. Schaffert, J., Cottin, M.C., Sonntag, A., Bobisch, C.A., Möller, R., Gauyacq, J.P., Lorente, N.: Tunneling electron induced rotation of a copper phthalocyanine moleculeon Cu(111). Phys. Rev. B **88**, 075410 (2013)

# Atom Manipulation Using Atomic Force Microscopy at Room Temperature

**Y. Sugimoto, M. Abe and S. Morita**

**Abstract** Atomic force microscopy (AFM) has demonstrated its capabilities as a nanotechnology tool. These capabilities include imaging/characterizing individual atoms on various surfaces and manipulating atoms and molecules. Here, we report how atom manipulation works on a well-known semiconducting surface, Si(111)-(7 × 7). To quantify the stochastic behavior of atom manipulation at room temperature (RT), atom hopping probabilities with various tip–surface distances are derived. The different hopping processes of Si adatoms have different tendencies in the probability plots. More remarkably, the ability of atom manipulation strongly depends on the AFM tip used. Tips can be characterized by their interaction force with surface Si adatoms. Force spectroscopic measurements combined with atom manipulation clarified that the ability to manipulate atoms is correlated with maximum attractive chemical bonding force with surface Si adatoms. Knowing the degree of chemical reactivity on the tip apex used for manipulation is key to enhancing the efficiency of the manipulation process occurring on semiconductor surfaces.

**Keywords** Atom manipulation · Non-contact atomic force microscopy · Force spectroscopy

Y. Sugimoto (✉)
Graduate School of Engineering, Osaka University, 2-1 Yamada-Oka, Suita, Osaka 565-0871, Japan
e-mail: sugimoto@afm.eei.eng.osaka-u.ac.jp

M. Abe
Graduate School of Engineering, Osaka University, 1-3 Machikaneyama-Cho, Toyonaka, Osaka 560-8531, Japan
e-mail: abe@stec.es.osaka-u.ac.jp

S. Morita
The Institute of Scientific and Industrial Research, Osaka University, 8-1 Mihogaoka, Ibaraki, Osaka 567-0047, Japan
e-mail: smorita@eei.eng.osaka-u.ac.jp



P. Moriarty and S. Gauthier (eds.), *Imaging and Manipulation of Adsorbates Using Dynamic Force Microscopy*, Advances in Atom and Single Molecule Machines,
DOI 10.1007/978-3-319-17401-3_3

# 1 Introduction

The ability to manipulate individual atoms on a surface is a remarkable feature of scanning probe microscopy [1]. Artificial nanostructures have been constructed for technological, physical, and chemical experiments using scanning tunneling microscopy (STM) at cryogenic temperatures [2–6]. Sophisticated atom positioning is usually controlled by the tiny force between the apex of the tip and the target atom on a surface [7].

Atomic force microscopy (AFM), the working principle of which is based on the detection of the tip–sample interaction force, has been developed as a versatile tool for imaging various surfaces, including insulator surfaces and organic molecular systems. The invention of a frequency modulation technique meant that atomic resolution could be routinely obtained [8]. The shift of frequency ($\Delta f$) from the resonance frequency ($f_0$) of an oscillated cantilever is detected for measurement of the tip–surface interaction force [9, 10]. One of the central analytical techniques in AFM is site-specific force spectroscopy measuring $\Delta f$ with the tip–surface distance [i.e., $\Delta f(z)$]. The interaction force between the atoms of the tip apex and individual surface atoms can be quantified by conversion from $\Delta f(z)$. Since the first demonstration of quantification of a single chemical bonding force on the Si(111)-(7 × 7) surface [11], various applications have used force spectroscopy with atomic resolution, including chemical identification by maximum attractive covalent bonding force [12, 13], direct estimation of magnetic exchange force [14, 15], and clarification of the fundamental relationship between current and chemical force [16, 17]. Extension of the dimensionality of the measured force field has been also developed. One can obtain a 3D force map by acquiring $\Delta f$ data in the volume on surfaces and converting it to a force map. There are many reports of force mapping measured on metal surfaces [18, 19], semiconductor surfaces [20–22], insulator surfaces [23–27], and molecules [28–33]. AFM has evolved from a surface imaging tool to a mapping tool for visualization of the force field on surfaces.

In addition to imaging and force spectroscopy, we have demonstrated that atom manipulation can be performed using AFM at large cantilever oscillation amplitudes [34–40]. Our first study of mechanical vertical manipulation using AFM was reported in 2003 [34]. We demonstrated that single Si adatoms can be extracted from the Si(111) surface reproducibly at low temperature (LT). Such vertical atom extraction is illustrated in Fig. 1a. A demonstration of lateral atom manipulation on a Ge surface at LT was published in 2005 [35]. Lateral manipulation means that single atoms residing on a surface are manipulated laterally (see Fig. 1d). Shortly, thereafter, we showed atom manipulation and assembly at room temperature (RT). This AFM manipulation achieved precision similar to that reported in LT manipulation using STM [36]. To avoid thermal diffusion at RT, we used a novel technique, called interchange lateral atom manipulation. In this method, the positions of two different elements are exchanged in the plane of a semiconducting alloy surface, as shown in Fig. 1e. In 2008, we discovered yet another interchange manipulation, called interchange vertical atom manipulation [38]. In this new

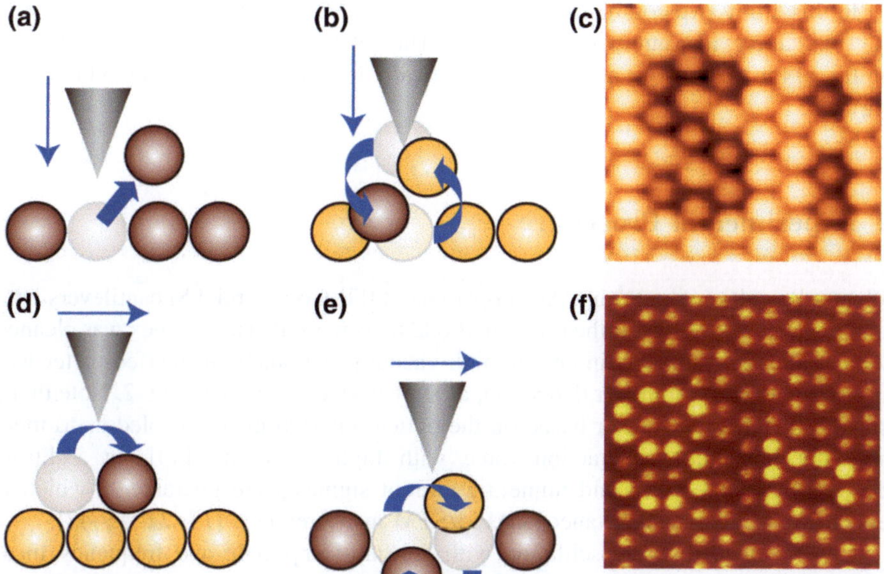


**Fig. 1** Four different types of atom manipulation and atom letters. **a** Illustration of vertical atom manipulation; **b** interchange vertical atom manipulation; **d** lateral atom manipulation; and **e** interchange lateral atom manipulation. **c** The symbol 'Si' assembled by interchange vertical manipulation at RT [38], and **f** the symbol 'Sn' assembled by interchange lateral manipulation at RT [36]

vertical manipulation, atoms on the tip apex are vertically exchanged with surface atoms as shown in Fig. 1b. This process is reproducible enough to engineer nanostructures on surfaces at RT. Interchange vertical/lateral atom manipulation allows us to construct planer nanostructures by rearranging two different adatom species on semiconductor surfaces at RT, as shown in Fig. 1c, f, respectively [36, 38]. These techniques at RT are highly promising for atom-by-atom construction of nanodevices able to operate in ambient conditions.

Moreover, using AFM, it is possible to investigate the interaction force associated with mechanical atom/molecule manipulation [18, 37, 38, 41–43]. Ternes et al. [18] measured the force required to move a metal adsorbate on a metallic surface using a quartz cantilever oscillated at sub-angstrom amplitude. Lateral manipulation of metal adsorbates is dominated by the lateral force only, independent of the vertical force. This result is in contrast to manipulation mechanisms proposed on the semiconducting surface where the vertical force plays an important role in atom movement by reducing the energy barrier [37]. The relation between mechanical atom manipulation and measured tip–surface interaction force, as well as the role played by the tip, is still open questions.

In this study, we combine the capability of AFM to manipulate single atoms at RT with force measurement techniques. The relation between tip reactivity and ability of the tip to move atoms is systematically investigated on the Si(111)-(7 × 7) surface.

## 2 Experimental Procedure

We used an ultrahigh vacuum AFM operated at RT. Commercial Si cantilevers with first resonant frequency of the order of 100 kHz were used. The tip apex was cleaned with an Ar ion sputtering in an ultrahigh vacuum. A homebuilt interferometer was used to detect the cantilever deflection, as shown schematically in Fig. 2. Note that a cantilever deflection sensor based on the optical interferometer enabled us to measure the tip–sample interaction force with high sensitivity [44]. In addition, decoupling of the force and tunneling current signals were guaranteed, which is essential for reliable simultaneous AFM/STM measurements [17, 22, 45–47].

Control of cantilever oscillation with constant amplitude and frequency measurements was performed using PLL-based commercial electronics. Cantilevers with a typical spring constant of 10–50 N/m were oscillated at a 10 nm order of oscillation amplitude to avoid jump into contact [48]. The tip–surface distance was regulated by feedback of the shift in the cantilever's first mechanical resonant frequency to the distance piezo. A commercial digital scanning controller was used for AFM scanning, force spectroscopic experiments, and atom manipulation experiments [49].

We used Si(111) samples of cut Si wafer for our atom manipulation and force spectroscopy experiments. The Si(111)-(7 × 7) surface was obtained by direct-current heating with vacuum pressure kept below $1 \times 10^{-10}$ Torr. The sample was flashed up to 1200 °C, followed by annealing.

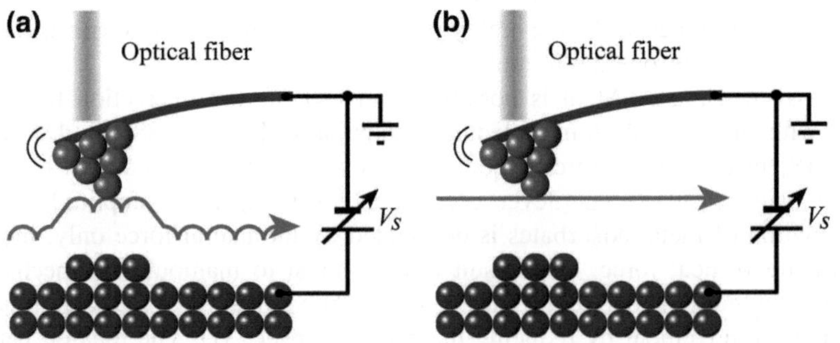


**Fig. 2** Schematic views of our cantilever-based AFM setup: **a** topographic scan mode; and **b** constant height scan mode

## 3 Manipulation of Si Atoms at Room Temperature

Although precise atom assembly was realized even at RT by the interchange lateral/vertical atom manipulation method, the manipulation mechanism on semiconducting surfaces at RT was not clear. Since the atom exchange process is complicated, we began by investigating lateral movement of the surface Si adatoms on the Si(111)-(7 × 7) surface by tip interaction. It was challenging to laterally manipulate single atoms that adsorbed on the surfaces at RT because of unwanted thermal diffusion. We found, however, that Si adatoms could be laterally manipulated toward a vacancy of adatom with high precision, even at RT [37, 39, 50]. Such vacant adatom positions were used as an open space toward which Si adatoms were manipulated.

Several AFM images of the same location on the Si surface are shown in Fig. 3a–h, which demonstrates manipulation of Si adatoms. The gray arrow in Fig. 3a indicates the vacancy created by vertical manipulation to supply a marker. An Si adatom vacancy was created by the mechanical atom extraction technique [34]. The procedure for this is given as follows. First, we scanned the Si(111)-(7 × 7) surface by AFM and selected a surface region that was suitable for atom manipulation experiments near the center of a terrace. The AFM scan was then stopped, and the tip was positioned over the Si adatom to be removed. After the feedback for the regulation of the tip–surface distance was deactivated, the tip approached the surface by a certain small displacement and then retracted to the original height. The feedback loop was then reactivated, and an AFM scan was carried out to check that atom extraction had been successful. The controlled indentation procedure was repeated with gradual increments of displacement until a Si adatom vacancy was successfully created. If the tip state became unstable after atom extraction, the tip was laterally moved to a different surface area, typically 10 nm away from the created vacancy site, and changed by intentional tip–surface contact or by AFM scanning under close tip–surface distances until the tip state was stabilized. When clear atomic resolution was obtained and the tip state was stable such that no further tip changes occurred during scanning, even under close tip–surface distances, the tip was brought to the area where the vacancy was created and an AFM topographic image was obtained over this region.

The white triangles in the images of Fig. 3a–h show an equivalent faulted half-unit cell, including the second vacancy created. In this triangle, Si adatoms were manipulated. The lateral manipulation of Si adatoms resulted in a change in the vacancy position, imaged as dark. The Si adatoms were manipulated along the dimer rows (along the sides of a triangle) in Fig. 3a–f. The atom could be also manipulated across the half-unit cell, as shown in Fig. 3f–h. In this manipulation, we were able to put a Si adatom on an unusual adsorption site (called an M site). An Si adatom adsorbed on the M site stably, such that the Si atom could be imaged by AFM within the experimental timescale under a weak tip–surface interaction. Figure 3g shows the AFM image of the artificial configuration. The vacancy is split, and three Si adatoms align horizontally in the half-unit cell. The atom configuration

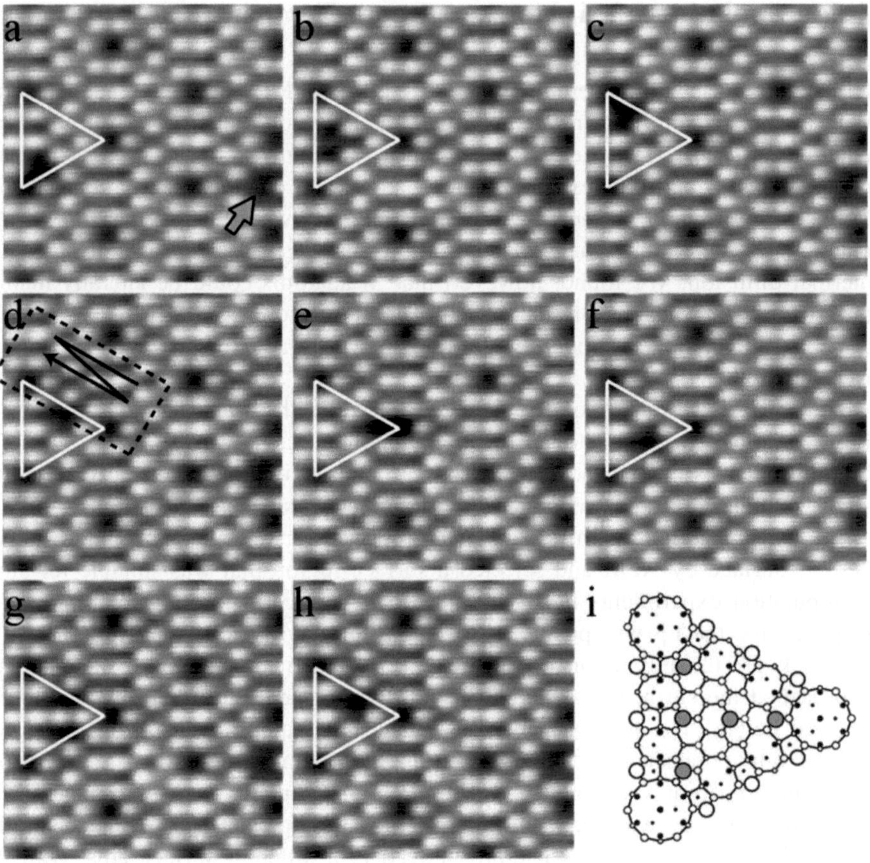


**Fig. 3** **a–h** AFM topographic images showing lateral manipulations of Si adatoms on the Si(111)-(7 × 7) [37]. The *gray arrow* in **a** indicates a missing adatom vacancy created by vertical manipulation for a marker. The acquisition parameters are $f_0$ = 162,295.8 Hz, $A$ = 282 Å, and $k$ = 28.7 N/m. Frequency shift set points for imaging ($\Delta f_{\text{image}}$) and manipulation ($\Delta f_{\text{move}}$) were −3.9 and −5.1 Hz, respectively. Si adatom can be put into an unusual metastable site, as shown in **i**, which is a stick and ball model that displays the atomic configuration

of this artificial structure is schematically illustrated in Fig. 3i. The manipulated Si adatom sits in a $T_4$ site near the center of the half-unit cell and forms bonds to an original surface rest atom.

These manipulations were performed using a vector scanning method in which the tip–sample distance feedback was closed, i.e., the topographic mode (see Fig. 2a). The procedure performed between Fig. 3d and e is shown in Fig. 3d. A Si adatom on a corner site (Co) was laterally manipulated to a center adatom site (Ce) by the method described below. After the initial image (Fig. 3d) was acquired, the scan area was reduced and the scan angle was rotated, as shown by the dotted rectangle. By the rotation of the scan angle, the fast scan direction coincided with the line above the Si adatom to be manipulated and the vacancy. The fast scan direction was set to be one way, indicated by the black arrow, and the slow scan started from the top right of the dotted rectangle. When the slow scan line reached the line that included the vacancy, it was stopped and scanning on the same line was repeated. The $\Delta f$ set point was then increased to be more negative, and the averaged tip–surface distance during line scans was decreased. Under a certain threshold distance, the Si adatom was manipulated toward the vacancy. After the signature of the atom movement appeared in the topographic signal, the averaged tip–surface distance was restored by setting the $\Delta f$ set point to the original value. The final image shown in Fig. 3e indicated that the Si atom had been successfully manipulated toward the vacancy.

These manipulations were categorized into the pulling mode. Si adatoms were moved by the attractive force (negative $\Delta f$). First principle calculations [37] shed light on the details of the mechanism. The chemical interaction force reduces the potential energy barrier, resulting in a thermal hop of the adatom toward an adjacent vacancy site. In contrast to typical lateral manipulation on metal surfaces [18], the vertical force plays a role in manipulation in our Si system. The tip pulls the target Si adatom up by the vertical component of the attractive force. Due to the vertical relaxation of Si adatoms, the bonds with the backbond atoms are weakened, which contributes to the barrier reduction. We believe that the proposed interpretation is characteristic of semiconductor surfaces.

## 4 Relationship Between Atom Manipulation and the Measured Force

To further quantify the manipulation process and investigate the effect of the reactivity of the tip apex on the manipulation, we combined force spectroscopy and constant height manipulation using the same tips. Before the lateral atom manipulation on the Si(111)-(7 × 7) surface, force spectroscopic measurements were carried out to characterize tip apex reactivity. We applied atom tracking to position the tip just above the adatom for site-specific force spectroscopy [51]. For quantitative atom manipulation experiments, it is necessary to precisely adjust the scan line just above the vacancy and the manipulated adatoms. To accomplish this at RT, the tip was first positioned above another Si adatom that was also above the line but in a different half-unit cell by atom tracking. After measurement of the drift velocity during atom tracking, thermal drift was compensated by the feedforward controller [20]. Successive line scans were then carried out above the vacancy and the manipulated Si adatoms in the [1 −1 0] direction (the left-right arrow), as illustrated in Fig. 4a. The scan direction was precisely aligned parallel to the [1 −1 0] direction by a versatile scan controller [49]. More importantly, the tip was scanned at constant heights by switching off the feedback control for the tip–surface distance to make statistics at each tip–surface distance. During constant height scanning, the

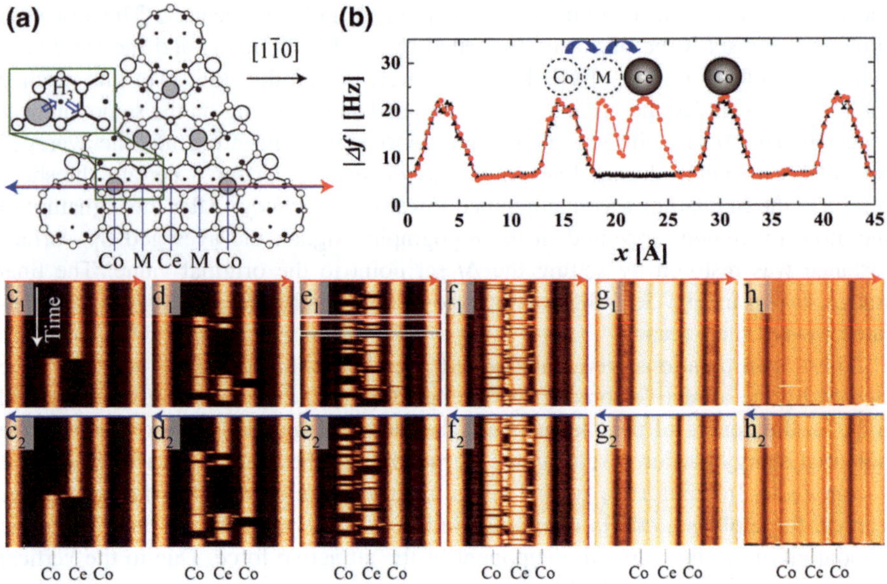


**Fig. 4** Quantitative manipulation of Si adatoms at RT [50]. **a** A ball and stick model of a half-unit cell in the Si(111)-(7 × 7) surface. Single vacancy is illustrated on a Ce site. The theoretically found atom hopping path (Co → $H_3$ → M) is shown in the *rectangle*. The *horizontal left* and *right arrows* indicate the direction of the vector tip scanning for manipulation at constant height. The tip was successively scanned above the line of the *arrows*, connecting a vacancy and manipulated adatoms. **b** Two selected line profiles of $\Delta f$ images from the constant height scan. **$c_1$–$h_2$** Constant height $\Delta f$ images at different tip–sample distances. The fast scan direction is from *left* to *right* ($c_1$–$h_1$) and *right* to *left* ($c_2$–$h_2$). Corresponding distances for the $\Delta f$ images: **$c_1$**, **$c_2$** $z = 0.21$ Å; **$d_1$**, **$d_2$** $z = 0.12$ Å; **$e_1$**, **$e_2$** $z = 0.02$ Å; **$f_1$**, **$f_2$** $z = -0.17$ Å; **$g_1$**, **$g_2$** $z = -0.31$ Å; and **$h_1$**, **$h_2$** $z = -1.16$ Å. Here, the $z$ values were defined from the force curves in Fig. 5a; i.e., $z = 0$ corresponding to maximum attractive force. Both line profiles in **b** were extracted from **$e_1$** and so were acquired with the same parameters. The profile with symbols of *black triangles* does not show the signature atom hopping. On the other hand, in the profile with *red circles*, we can see the signature atom hop from a *left* Co site to an M site, followed by a successive jump to a Ce site

$\Delta f$ signal was recorded (see Fig. 2b). This scanning mode is indispensable for reducing the risk of a tip crash during atom manipulation. The slope of the $\Delta f$ ($z$) curves above the adatom sites changes the polarity at the tip–surface distances where atom hopping probability is large, because the long-range force is relatively small for sharp Si tips.

Figure 4$c_1$–$h_2$ shows the $\Delta f$ images of the constant height line scans on the Si (111)-(7 × 7) surface. The horizontal arrows in Fig. 4a indicate the scan line. Each image in Fig. 4$c_1$–$h_2$ shows the $\Delta f$ signal at a constant height vector scan, so the bright stripes correspond to individual Si adatoms. Discontinuity and intermittent patterns reflect atom hopping processes caused by interaction with the tip. The fast scan direction from left to right is shown in ($c_1$–$h_1$) and right to left is shown in ($c_2$–$h_2$). All images in Fig. 4 were acquired using the same tip. From ($c_1$, $c_2$) to ($h_1$, $h_2$), the tip–surface distance decreased: $\Delta f$ images in $c_1$ and $c_2$ were measured at a far tip–surface distance, while those in $h_1$ and $h_2$ were measured at the closest tip–surface distance.

The Si adatoms were moved following the tip scan when the tip–surface distance became small. From the $\Delta f$ images, we can see which adsorption sites were occupied by the Si adatoms, so atom movement could be followed. The atom manipulation process is stochastic at RT. Different manipulation processes could be observed in line profiles of $\Delta f$ images even at the same scan parameter of the tip–sample distance [39]. For example, two different line profiles are shown in Fig. 4b. These line profiles were taken from the $\Delta f$ image as shown in Fig. 4e$_1$. The line profile with the black triangles shows that two Si adatoms stay at left and right Co sites without any atom movement. In contrast, the line profile with the red circles shows the signature of atom displacement. Initial atom configuration is the same between the two profiles. The latter profile demonstrates atom hopping as follows. When the tip scans from the left, the left Co adatom is imaged. At $x$ = 17.8 Å, the Si adatom is moved to the M site following the tip, causing the $\Delta f$ jump. When the tip reaches $x$ = 20.7 Å, the same Si adatom jumps to the Ce site. There was a vacancy at the Ce site before this manipulation, while the vacancy appears on the left Co site after the scan.

Figure 4b shows that the manipulation process is stochastic at RT. To quantify the probability of successful manipulation, we analyzed a number of $\Delta f$ images measured at various tip–surface distances. We calculated the probability as the number of line profiles showing particular manipulation processes divided by the number of total line profiles. The results obtained at a certain tip state are shown in Fig. 5a. Three different probability plots are shown as a function of the tip–surface distance. Solid (open) circles show the probabilities of atom hops from left (right) Co to M sites, and open triangles show the probabilities of hops from Ce to M sites. In Fig. 5a, the chemical bonding force [$F_z(z)$] and chemical bonding potential [$U(z)$] are shown together. Those curves were numerically converted from the measured $\Delta f(z)$ curve above the Si adatom by the same tip state used for manipulation experiments. The tip approach increases the probability of atom hopping for all three processes. This increase was not step-like as expected at absolute zero temperature. Around the maximum $F_z(z)$, the probabilities smoothly increase from 0 to 100 %, which is a feature of manipulation at RT.

Most remarkably, the success rate of the manipulation strongly depended on the tip apex state. To systematically investigate the tip dependence of the manipulation, we carried out the same experiments with 15 different tips using 10 different cantilevers. Different tip states for the same cantilever were prepared by controlled tip crash to the surface. In our extended studies, two kinds of tips were found, i.e., the tips that could move atoms and those that could not. Figure 5a is an example of the former group, while Fig. 5b is an example of the latter. The probability of atom movement remains zero even for tip–surface distances closer than the distance of maximum attractive force. Interestingly, the tip state that does not have manipulation capability can be changed. After the intentional tip change, the results shown in Fig. 5c were obtained. This tip state has the ability of atom manipulation. All

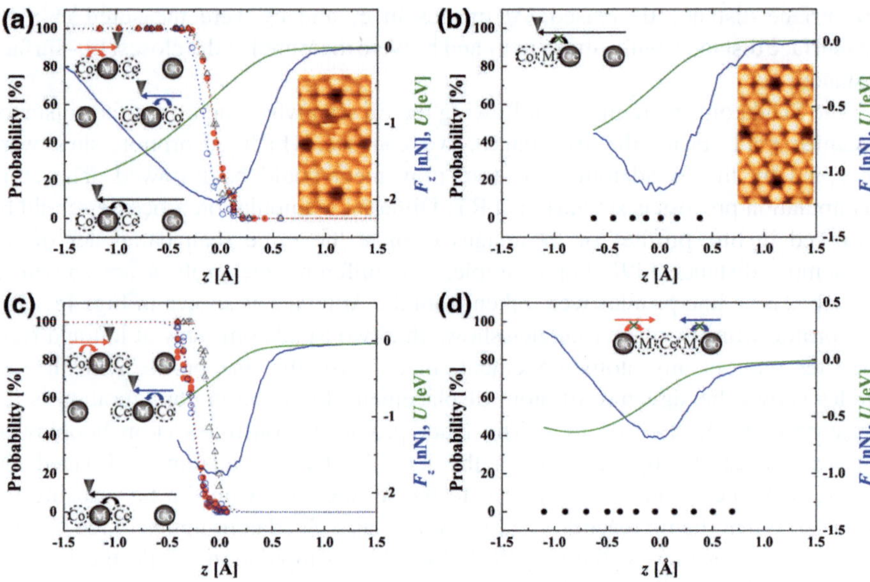


**Fig. 5** Hopping probability plots with $F_z(z)$ and $U(z)$ curves. Results with four different tips are shown [50]. **a** *Solid circles* correspond to the probability of the atom hopping from the left Co to the M site. *Open circles* correspond to the likelihood of the atom hopping from the right Co to the M site. *Open triangle* shows Ce to the M site. **b** This data do not show atom manipulation. **c** The data set acquired by using the same cantilever as in **b**, but with a different tip state. The different tip state is prepared by a gentle tip–surface crash. Assignment of the symbols for the manipulation process is the same as in **a**. **d** These data do not show atom manipulation. The *insets* of **a** and **b** show corresponding AFM topographic images

processes observed in Fig. 5a were reproduced in Fig. 5c. The different tendencies of the manipulation between Fig. 5a and c will be discussed later. Figure 5d shows the result with a different tip state and different cantilever from those used in Figs. 5a–c. In Fig. 5d, the Si adatoms cannot be moved in either the forward or backward scan direction even at distances where $F_z$ on Si adatom reaches the repulsive force region. These results clearly show that the properties of the few atoms at the very apex of the tip are essential to manipulation.

We found that manipulation capability can be characterized by the interaction force with surface adatoms. Figure 6a displays a histogram of magnitude of the maximum attractive force $F_{max}$ of 15 different tips. The forces were measured above the Si adatom. In the graph, it is clear that manipulation capability is significantly correlated with $F_{max}$. Thus, $F_{max}$ can be used to categorize the 15 tips measured into two groups. The group with larger $F_{max}$ has the capability to manipulate Si adatoms, while it is not possible to move atoms using the group with the smaller $F_{max}$. We call the tips with larger $F_{max}$ reactive tips and the others less-reactive tips. To manipulate Si adatoms as shown in Fig. 3, reactive tips are necessary.

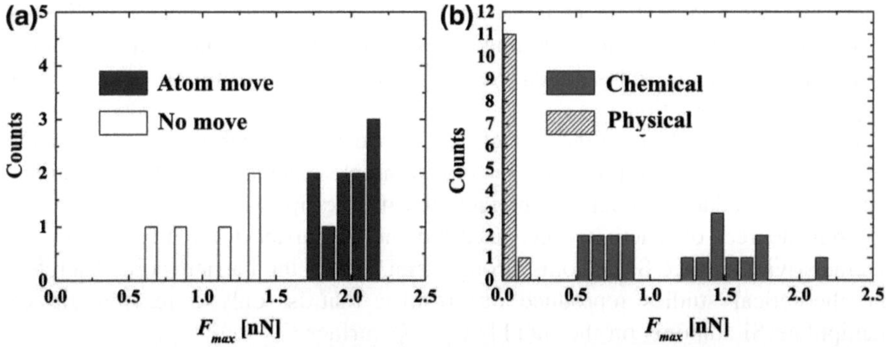


**Fig. 6** **a** The histogram of the maximum attractive forces with 15 different tips [50]. The forces were acquired by force spectroscopy above Si adatoms. There are two groups. The tips with larger attractive force can move atoms. The other tips with smaller force cannot move atoms even at close tip–surface distances. **b** The histogram of maximum attractive short-range forces measured above Si adatoms besides atom manipulation experiments. There is a group showing tiny $F_{\max}$, i.e., physical force [52]

These two types of tips can be identified from AFM images with relatively close tip–surface distances. With a reactive tip, during AFM scanning, one will be able to observe that Si adatoms are moved following the tip (see the inset image in Fig. 5a). In contrast, with a less-reactive tip, one will obtain an AFM image without fluctuation around the vacancy site. In Fig. 5b, even Si rest atoms can be clearly resolved.

It is worth nothing that there is a third class of tip apex, i.e., the non-reactive tip. Non-reactive tips cannot provide clear atomic resolution on the Si(111)-(7 × 7) surfaces by AFM (as opposed to all the tips in Fig. 6a that gave a clear atomic resolution). In addition to the atom manipulation experiments shown in Fig. 6a, we measured $F_{\max}$ on Si adatoms on the Si(111)-(7 × 7) surfaces using various tips. As shown in the histogram of Fig. 6b, the tips that cannot provide atomic resolution have extremely small $F_{\max}$, usually less than 0.1 nN [52]. The other tips giving atomic resolution have $F_{\max}$ larger than 0.5 nN, and the dispersion of $F_{\max}$ is the same as that in Fig. 6a.

Thus, it is possible to manipulate Si adatoms with reactive tips. More specifically, we can discuss the anisotropy of the manipulation direction depending on the tip. Direction can be investigated by equivalent manipulation processes from the Co to the M site, but in opposite scan directions. In Fig. 5a, solid (open) circles indicate the probability plots from the left (right) Co site to the M site. Although those processes are equivalent, the probability plots differ with the scan directions. The probability from left to right starts to increase at farther distances than that from right to left. This is likely due to a tip asymmetry effect. Asymmetric tip structures may also produce asymmetric dangling bond states with respect to the surface. Recently, Jarvis et al. [53] reported theoretical studies about tip effects on vertical manipulation. The manipulation depended on the orientation of the dangling bond.

Therefore, we can assume that the tip had an asymmetric structure in Fig. 5a. Symmetric results can be also obtained, as shown in Fig. 5c. There, the probability plots for different tip scans match. Such a tip may have symmetric structure with respect to scan direction.

Finally, we should note that correlation between the capability of atom manipulation and $F_{max}$ was reproduced by first principle calculations [50]. The energy barrier for Si adatom under the interaction with the tip was calculated. Tips with different degrees of reactivity decreased the energy barrier differently. The tip that interacts with surface Si adatoms strongly can reduce the barrier more. Therefore, the theoretical studies reproduce our results; that is, only a reactive tip can manipulate Si adatoms on the Si(111)-(7 × 7) surface.

## 5 Conclusions

We performed force spectroscopic measurements on Si adatoms on Si(111)-(7 × 7) surfaces using various tips of AFM. It was found that tips could be categorized into three groups: non-reactive, less reactive, and more reactive. Non-reactive tips interact with the adatom by physical force, and atomic resolution cannot be clearly obtained on the surface by AFM. The other two groups can give clear atomic resolution by chemical interaction force. Using the latter two classes of tips, we systematically investigated the relation between degree of reactivity on AFM tip and capability of lateral manipulation on the Si(111)-(7 × 7) surface at RT. It was found that the less-reactive tips did not have the capability of atom manipulation, while the more-reactive tips did. We found clear correlation between the capability of the manipulation and the degree of tip reactivity that can be characterized by the magnitude of the maximum attractive force with surface Si adatoms. This may be related to the 'sticky fingers problem' pointed out by Smalley [54]. For efficient manipulation, one should tune the extent of the chemical reactivity with the interaction strength between manipulated atom and substrate in mind. Once a suitable tip is obtained, it will be possible to manipulate atoms at will—even on reactive semiconducting surfaces—by adjusting the tip–surface interaction through the distance control.

**Acknowledgments** This work was supported by a grant-in-aid for Scientific Research from MEXT, Funding Program for Next Generation World-Leading Researchers. I would like to thank my colleagues who contributed to this study: Ayhan Yurtsever, Oscar Custance, Martin Ondracek, Pavel Jelinek, Pablo Pou, and Ruben Perez.

## References

1. Eigler, D.M., Lutz, C.P., Rudge, W.E.: Nature (London) **352**, 600 (1991)
2. Nilius, N., Wallis, T.M., Ho, W.: Science **297**, 1853–1856 (2002)
3. Heinrich, A.J., Lutz, C.P., Gupta, J.A., Eigler, D.M.: Science **298**, 1381 (2002)

4. Hla, S.W., Braun, K.F., Rieder, K.H.: Phys. Rev. B **67**, 201402 (2003)
5. Loth, S., Baumann, S., Lutz, C.P., Eigler, D.M., Heinrich, A.J.: Science **335**, 196–199 (2012)
6. Khajetoorians, A.A., et al.: Science **339**, 55–59 (2013)
7. Bartels, L., Meyer, G., Rieder, K.-H.: Phys. Rev. Lett. **79**, 697 (1997)
8. Albrecht, T.R., Grütter, P., Horne, D., Rugar, D.: J. Appl. Phys. **69**, 668–673 (1991)
9. Morita, S., Wiesendanger, R., Meyer, E. (eds.): Noncontact Atomic Force Microscopy. Springer, Berlin (2002)
10. Giessibl, F.J.: Rev. Mod. Phys. **75**, 949–983 (2003)
11. Lantz, M.A., et al.: Science **291**, 2580–2583 (2001)
12. Sugimoto, Y., et al.: Nature (London) **446**, 64–67 (2007)
13. Setvin, M., et al.: ACS Nano **6**, 6969–6976 (2012)
14. Schmidt, R., et al.: Phys. Rev. Lett. **106**, 257202 (2011)
15. Pielmeier, F., Giessibl, F.J.: Phys. Rev. Lett. **110**, 266101 (2013)
16. Ternes, M., et al.: Phys. Rev. Lett. **106**, 016802 (2011)
17. Sugimoto, Y., et al.: Phys. Rev. Lett. **111**, 106803 (2013)
18. Ternes, M., Lutz, C.P., Hirjibehedin, C.F., Giessibl, F.J., Heinrich, A.J.: Science **319**, 1066–1069 (2008)
19. Albers, B.J., et al.: Nat. Nanotechnol. **4**, 307 (2009)
20. Abe, M., et al.: Appl. Phys. Lett. **90**, 203103 (2007)
21. Sugimoto, Y., Namikawa, T., Miki, K., Abe, M., Morita, S.: Phys. Rev. B **77**, 195424 (2008)
22. Sugimoto, Y., Ueda, K., Abe, M., Morita, S.: J. Phys.: Condens. Matter **24**, 084008 (2012)
23. Holscher, H., Langkat, S.M., Schwarz, A., Wiesendanger, R.: Appl. Phys. Lett. **81**, 4428 (2002)
24. Schirmeisen, A., Weiner, D., Fuchs, H.: Phys. Rev. Lett. **97**, 136101 (2006)
25. Ruschmeier, K., Schirmeisen, A., Hoffmann, R.: Phys. Rev. Lett. **101**, 156102 (2008)
26. Kawai, S., Glatzel, T., Koch, S., Baratoff, A., Meyer, E.: Phys. Rev. B **83**, 035421 (2011)
27. Rahe, P., et al.: Rev. Sci. Instrum. **82**, 063704 (2011)
28. Ashino, M., Schwarz, A., Behnke, T., Wiesendanger, R.: Phys. Rev. Lett. **93**, 136101 (2004)
29. Ashino, M., et al.: Nat. Nanotechnol. **3**, 337 (2008)
30. Gross, L., Mohn, F., Moll, N., Liljeroth, P., Meyer, G.: *Science* **325**, 1110–1114 (2009)
31. Braun, D.A., Weiner, D., Such, B., Fuchs, H., Schirmeisen, A.: Nanotechnology **20**, 264004 (2009)
32. Sun, Z., Boneschanscher, M.P., Swart, I., Vanmaekelbergh, D., Liljeroth, P.: Phys. Rev. Lett. **106**, 046104 (2011)
33. Mohn, F., Gross, L., Meyer, G.: Appl. Phys. Lett. **99**, 053106 (2011)
34. Oyabu, N., Custance, O., Yi, I., Sugawara, Y., Morita, S.: Phys. Rev. Lett. **90**, 176102 (2003)
35. Oyabu, N., Sugimoto, Y., Abe, M., Custance, O., Morita, S.: Nanotechnology **16**, S112 (2005)
36. Sugimoto, Y., et al.: Nat. Mater. **4**, 156–159 (2005)
37. Sugimoto, Y., et al.: Phys. Rev. Lett. **98**, 106104 (2007)
38. Sugimoto, Y., et al.: Science **322**, 413–417 (2008)
39. Sugimoto, Y., Miki, K., Abe, M., Morita, S.: Phys. Rev. B **78**, 205305 (2008)
40. Sugimoto, Y., Yurtsever, A., Hirayama, N., Abe, M., Morita, S.: Nat. Commun. **5**, 4360 (2014)
41. Sweetman, A., et al.: Phys. Rev. Lett. **106**, 136101 (2011)
42. Mao, H.Q., Li, N., Chen, X., Xue, Q.K.: J. Phys.: Condens. Matter **24**, 084004 (2012)
43. Langewisch, G., Falter, J., Fuchs, H., Schirmeisen, A.: Phys. Rev. Lett. **110**, 036101 (2013)
44. Morita, K., Sugimoto, Y., Sasagawa, Y., Abe, M., Morita, S.: Nanotechnology **21**, 305704 (2010)
45. Sawada, D., Sugimoto, Y., Morita, K., Abe, M., Morita, S.: Appl. Phys. Lett. **94**, 173117 (2009)
46. Sugimoto, Y., Yi, I., Morita, K., Abe, M., Morita, S.: Appl. Phys. Lett. **96**, 263114 (2010)
47. Sugimoto, Y., et al.: Phys. Rev. B **81**, 245322 (2010)
48. Giessibl, F.J., Hembacher, S., Herz, M., Schiller, C., Mannhart, J.: Nanotechnology **15**, S79 (2004)

49. Horcas, I., et al.: Rev. Sci. Instrum. **78**, 013705 (2007)
50. Sugimoto, Y., et al.: ACS Nano **7**, 7370–7376 (2013)
51. Abe, M., Sugimoto, Y., Custance, O., Morita, S.: Appl. Phys. Lett. **87**, 173503 (2005)
52. Yurtsever, A., et al.: Phys. Rev. B **87**, 155403 (2013)
53. Jarvis, S., Sweetman, A., Bamidele, J., Kantorovich, L., Moriarty, P.: Phys. Rev. B **85**, 235305 (2012)
54. Smalley, R.E.: Sci. Am. **285**, 76 (2001)

# A Considered Approach to Force Extraction from Dynamic Force Microscopy Measurements

**Andrew Stannard and Adam M. Sweetman**

**Abstract** High-accuracy measurements of short-range chemical interactions between well-defined atomic and molecular species are possible using dynamic force microscopy (DFM). For these measurements to be quantitatively reliable, proper attention must be paid to the analysis techniques used to process experimental data; a considered approach is detailed here. We describe a rigorous implementation of the 'on–off' method to remove long-range interactions and quantify the uncertainty arising from implementation of this technique. We also explore the validity of approximation formulae used to convert from frequency shift to force, finding that these approximations are only valid when the oscillation amplitude is at least an order of magnitude different than any characteristic interaction length scale.

**Keywords** Dynamic force microscopy · Force conversion · Chemical interactions

## 1 Introduction

In dynamic force microscopy (DFM), a cantilever of stiffness $k$ with a sharp tip attached to its free end is driven to oscillate at its resonant frequency $f_0$ with amplitude $a$. When the tip is brought into close proximity to a sample surface, tip–sample interactions cause a shift in the cantilever's resonant frequency $\Delta f(z)$, dependent on the tip–sample separation $z$ (defined throughout here as the separation at the point of closest approach during an oscillation cycle). In DFM, the term 'spectroscopy' is often used to describe the measurements of frequency shift obtained whilst the tip–sample

A. Stannard (✉) · A.M. Sweetman
School of Physics and Astronomy, University of Nottingham,
Nottingham NG7 2RD, UK
e-mail: andrew.stannard@nottingham.ac.uk

A.M. Sweetman
e-mail: adam.sweetman@nottingham.ac.uk



P. Moriarty and S. Gauthier (eds.), *Imaging and Manipulation of Adsorbates Using Dynamic Force Microscopy*, Advances in Atom and Single Molecule Machines,
DOI 10.1007/978-3-319-17401-3_4

separation is controllably decreased (or increased). This spectroscopy data can then be processed to extract the corresponding tip–sample interaction force $F(z)$—a thorough description of this analysis procedure is presented here.

DFM is commonly used to obtain high-accuracy measurements of short-range chemical interactions between tip- and surface-adsorbed species. These measurements are highly sensitive to the processing procedure, resulting in unreliable quantitative values if proper care and attention is not taken. Two obvious sources of error are the calibration uncertainties of cantilever stiffness and oscillation amplitude. However, even data obtained from a perfectly characterized and calibrated system require a thorough and considered analysis procedure to convert measured frequency shift data into the desired short-range interaction force. This procedure can be separated into two stages—subtraction of long-range contributions and conversion from frequency shift to force (typically in this order). Without proper consideration of these stages, errors at, or greater than, the magnitude of inherent calibration uncertainties can be introduced.

In the following, we first discuss the formulae used to convert frequency shift into force, and using a prototypical interaction potential, simplifying approximation formulae are tested to ascertain under which conditions they are applicable. Second, a rigorous method for the isolation of short-range forces via the subtraction of long-range forces is explained in detail, and the uncertainties introduced via such a process are divulged in a transparent manner.

## 2 Relationship Between Frequency Shift and Force

In DFM, it is generally the tip–sample interaction force, $F(z)$, that is of most interest. However, knowledge of the tip–sample interaction potential, $U(z)$, can also be informative. These two are (naturally) related via $F(z) = -\mathrm{d}U(z)/\mathrm{d}z$ and $U(z) = \int_z^\infty F(t)\,\mathrm{d}t$, where in the latter, $t$ is used as the integration variable representing tip–sample separations to avoid using $z$ as both the integration variable and the lower bound of the integral ($z$ represents the tip–sample separation at which the interaction potential is to be determined).

A variety of formulae for converting cantilever frequency shift, $\Delta f(z)$, to tip–sample interaction force (and vice versa) are available. The applicability of each formula depends on the magnitude of the oscillation amplitude compared to characteristic length scales of the interaction being probed. This section discusses these different formulae and demonstrates when they can—and, more importantly, cannot—be employed successfully.

It should be noted that conversion from frequency shift to force can also be achieved via a matrix method proposed by Giessibl [1]. Reference [2] compares this matrix method to the formula-based approach (to be discussed here) proposed by Sader and Jarvis [3].

## 2.1 Arbitrary Oscillation Amplitudes

For oscillation amplitudes of an arbitrary magnitude, the frequency shift at the point of closest approach relates to an integration of the interaction force over the tip–sample separation interval $[z, z + 2a]$ (the range spanned during an oscillation cycle) and may be written as

$$\Delta f(z) = -\frac{f_0}{\pi k a} \int_0^{\pi} F(z + a(1 + \cos\phi)) \cos\phi \, \mathrm{d}\phi, \tag{1}$$

where $\phi$ represents the cantilever oscillation phase, with $\phi = \pi$ at the point of closest approach [4]. It is very important to note that the amplitude, $a$, refers to the maximum displacement of the tip from its equilibrium position during an oscillation cycle and not the peak-to-peak displacement separation. Other manifestations of Eq. 1 can be found in the literature (see, for example Ref. [3]) for an improper integral version that may be converted to Eq. 1 by a change of variable.

It is informative to study the oscillation amplitude dependence of frequency shift as a function of tip–sample separation for a prototypical DFM interaction, namely the chemical bond formation between a silicon-terminated tip and a silicon adatom (we ignore background interactions at this stage, justified by the fact that additional interactions only change frequency shifts additively). The Morse potential model of chemical bonding can be expressed as

$$U_{\mathrm{Morse}}(z) = U_{\mathrm{e}}\left(\mathrm{e}^{-2(z-z_{\mathrm{e}})/\lambda} - 2\mathrm{e}^{-(z-z_{\mathrm{e}})/\lambda}\right), \tag{2}$$

where $U_{\mathrm{e}}$ is the equilibrium binding energy, $z_{\mathrm{e}}$ is the equilibrium bond length, and $\lambda$ is a characteristic length scale of the potential. For the system of interest, Pérez et al. [5] fit a Morse potential to density functional theory (DFT) data of a silicon-terminated tip interacting with a silicon adatom and obtain $U_{\mathrm{e}} = 2.27$ eV, $z_{\mathrm{e}} = 235$ pm, and $\lambda = 78.7$ pm. Using Eq. 1, and the corresponding 'Morse force', the cantilever frequency shift behaviour can be calculated for this prototypical interaction for a range of oscillation amplitudes (here using $f_0 = 25$ kHz and $k = 2.2$ kN/m, typical parameters for a qPlus sensor). Figure 1a shows frequency shift curves for amplitudes spanning four orders of magnitude—1 pm to 10 nm. It is clear that increasing the amplitude reduces frequency shift magnitudes and moves the minimum value towards smaller tip–sample separations.

Sader and Jarvis [3] derived an expression for how the interaction force at the point of closest approach relates to an integration of the frequency shift (and frequency shift gradient) over the tip–sample separation interval $[z, \infty]$,

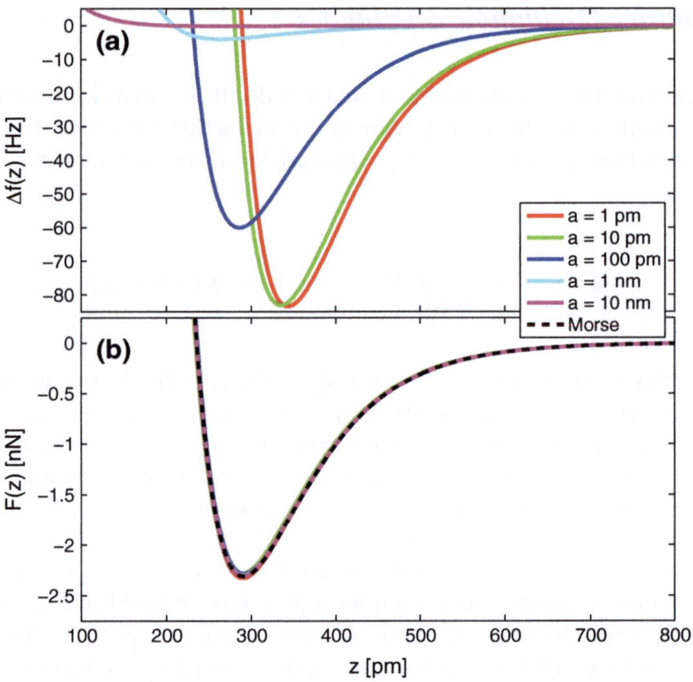


**Fig. 1** Calculations applied to the Morse potential of a silicon-terminated tip interacting with a silicon adatom. **a** Frequency shift curves display a strong dependence on oscillation amplitude, increasing the amplitude reduces the magnitude and moves the minimum value to smaller separations. **b** Reconverted interaction forces for all amplitudes show excellent agreement with the Morse force

$$F(z) = \frac{2k}{f_0}\int_z^{\infty}\left(1 + \frac{a^{1/2}}{8\sqrt{\pi(t-z)}} - \frac{a^{3/2}}{\sqrt{2(t-z)}}\frac{\mathrm{d}}{\mathrm{d}t}\right)\Delta f(t)\,\mathrm{d}t, \tag{3}$$

with the corresponding interaction potential given by

$$U(z) = \frac{2k}{f_0}\int_z^{\infty}\left((t-z) + \frac{a^{1/2}}{4}\sqrt{\frac{(t-z)}{\pi}} + \frac{a^{3/2}}{\sqrt{2(t-z)}}\right)\Delta f(t)\,\mathrm{d}t. \tag{4}$$

To demonstrate the high accuracy of Sader and Jarvis' formulae, the frequency shift curves shown in Fig. 1a are reconverted to interaction force using Eq. 3. The results are shown in Fig. 1b; clearly, there is excellent agreement with the original Morse form. As a quotable value of accuracy, the error in the recovered maximum attractive forces is less than 2 % across all amplitudes (the equivalent comparison of interaction potentials is equally as accurate).

## 2.2 Small Oscillation Amplitudes

For sufficiently small oscillation amplitudes, $\Delta f(z)$ converges to an amplitude-independent form. The small-amplitude approximation assumes that the interaction force gradient is constant over an oscillation cycle. Under this approximation, the frequency shift is related to the interaction force gradient by

$$\Delta f_{\mathrm{small}}(z) = -\frac{f_0}{2k}\frac{\mathrm{d}F(z)}{\mathrm{d}z}. \tag{5}$$

as discussed by Albrecht et al. [6]. The interaction force at the point of closest approach relates to an integration of the frequency shift over the tip–sample separation interval $[z, \infty]$,

$$F_{\mathrm{small}}(z) = \frac{2k}{f_0}\int_z^{\infty} \Delta f(t)\,\mathrm{d}t, \tag{6}$$

with the corresponding interaction potential given by

$$U_{\mathrm{small}}(z) = \frac{2k}{f_0}\int_z^{\infty} (t - z)\Delta f(t)\,\mathrm{d}t. \tag{7}$$

It can be seen that Eqs. 6 and 7 correspond to the first terms inside the brackets of Eqs. 3 and 4, respectively.

It is informative to consider under which practical circumstances the small-amplitude approximation is valid, i.e. what qualifies as a 'small' oscillation amplitude for typical experimentally probed interactions. Returning to the Morse potential used previously, Fig. 2a shows a comparison of $\Delta f(z)$ calculated using the small-amplitude approximation (Eq. 5) to equivalent curves obtained at a range of amplitudes with no approximation (Eq. 1). It can be seen that $a$ = 1 pm agrees excellently, $a$ = 10 pm agrees satisfactorily, but $a$ = 100 pm quite strongly disagrees, and $a$ = 1 nm even more so.

In general, it is of more interest to know how well the small-amplitude approximation works for converting frequency shift into interaction force. As such, we can take the amplitude-dependent frequency shift curves from Fig. 2a and use the small-amplitude approximation (Eq. 6) to reconvert to interaction forces. The results are shown in Fig. 2b. One observes similar behaviour as with the frequency shifts, $a$ = 1 pm agrees excellently, $a$ = 10 pm agrees satisfactorily, but the error in the reconverted maximum attractive force is approximately 30 % for $a$ = 100 pm, an unsatisfactorily high value, and only gets worse with increasing amplitude (comparisons of interaction potential display similar behaviour). Overall, it is clear that for this particular Morse potential, the small-amplitude approximation can only be successfully employed for $a \lesssim 10\,\mathrm{pm}$.

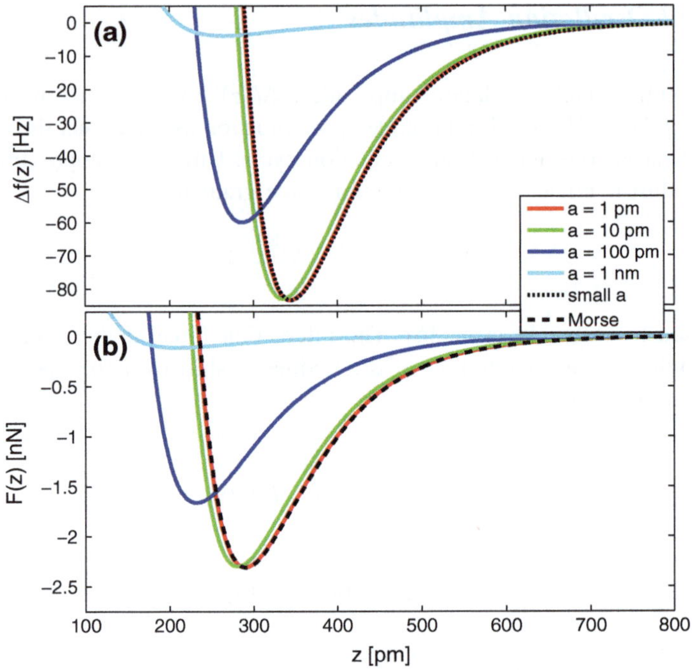


**Fig. 2** Calculations applied to the Morse potential of a silicon-terminated tip interacting with a silicon adatom, made using the small-amplitude approximation. Good agreement is displayed for $a \lesssim 10$ pm in both **a** frequency shifts and **b** recovered tip–sample forces

## *2.3 Large Oscillation Amplitudes*

For a given tip–sample interaction, increasing the oscillation amplitude always reduces the frequency shift magnitude. However, for sufficiently large amplitudes, $\Delta f(z) \cdot a^{3/2}$ converges to an amplitude-independent form. The large-amplitude approximation assumes that tip–sample interactions near the point of closest approach dominate [7]. Under this approximation, the frequency shift at the point of closest approach is related to an integration of the interaction force over the tip–sample separation interval $[z, z + 2a]$,

$$\Delta f_{\mathrm{large}}(z) = \frac{f_0}{\sqrt{2}\pi k a^{3/2}} \int_{z}^{z+2a} \frac{F(t)}{\sqrt{t-z}}\,\mathrm{d}t. \tag{8}$$

The interaction force at the point of closest approach is related to an integration of the frequency shift gradient over the tip–sample separation interval $[z, \infty]$,

$$F_{\text{large}}(z) = -\frac{\sqrt{2}ka^{3/2}}{f_0}\int_z^{\infty}\frac{1}{\sqrt{t-z}}\frac{\mathrm{d}\Delta f(t)}{\mathrm{d}t}\,\mathrm{d}t, \tag{9}$$

as shown by Dürig [8], with the corresponding interaction potential given by

$$U_{\text{large}}(z) = \frac{\sqrt{2}ka^{3/2}}{f_0}\int_z^{\infty}\frac{\Delta f(t)}{\sqrt{t-z}}\,\mathrm{d}t. \tag{10}$$

It can be seen that Eqs. 9 and 10 correspond to the third terms inside the brackets of Eqs. 3 and 4, respectively.

It is informative to consider under which practical circumstances the large-amplitude approximation is valid, i.e. what qualifies as a 'large' oscillation amplitude for typical experimentally probed interactions. Returning to the Morse potential used previously, Fig. 3a shows a comparison of $\Delta f(z) \cdot a^{3/2}$ calculated using the large-amplitude approximation (Eq. 8) to equivalent curves obtained at a range of large amplitudes with no approximation (Eq. 1). It can be seen that

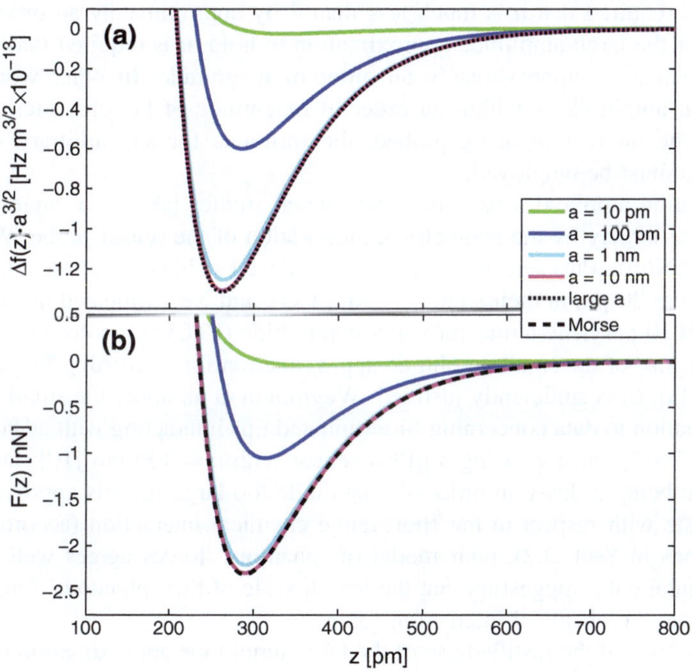


**Fig. 3** Calculations applied to the Morse potential of a silicon-terminated tip interacting with a silicon adatom, made using the large-amplitude approximation. Good agreement is displayed for $a \gtrsim 1$ nm in both **a** amplitude-modified frequency shifts $\Delta f(z) \cdot a^{3/2}$ and **b** recovered tip–sample forces

$a$ = 10 nm agrees excellently, $a$ = 1 nm agrees satisfactorily, but $a$ = 100 pm quite strongly disagrees, and $a$ = 10 pm even more so.

Again, it is of more interest to know how well the large-amplitude approximation works for converting frequency shift into interaction force. As such, we can take the frequency shift curves from Fig. 3a and use the large-amplitude approximation (Eq. 9) to reconvert to interaction force. The results are shown in Fig. 3b. Again, one observes similar behaviour as with the frequency shifts; $a$ = 10 nm agrees excellently, $a$ = 1 nm agrees satisfactorily, but the error in the reconverted maximum attractive force is approximately 55 % for $a$ = 100 pm, an unacceptably high value, and only gets worse with decreasing amplitude (comparisons of interaction potential display similar behaviour). Overall, it is clear that for this particular Morse potential, the large-amplitude approximation can only be successfully employed for $a \gtrsim 1$ nm.

## 2.4 Discussion on the Use of Approximation Formulae

It appears that there is a general rule of thumb here relating the oscillation amplitude, $a$, and the characteristic length scale of the interaction, $\lambda$ ($\approx$100 pm for the Morse potential used throughout here). For the small-amplitude approximation to hold, it is required that $a$ is much less than $\lambda$ by approximately an order of magnitude; for the large-amplitude approximation to hold, it is required that $a$ is much greater than $\lambda$ by approximately an order of magnitude. In other words, if the oscillation amplitude is within an order of magnitude of the characteristic length scale of the interaction being probed, the formulae for any arbitrary oscillation amplitude must be employed.

Oscillation amplitudes used in experimental studies take on a range of values, determined largely by the geometric configuration of the sensor probe. With a stiff ($k \simeq 1.8$ kN/m) qPlus sensor, Kichin et al. achieved stable oscillation amplitudes as small as $a \simeq 20$ pm in their studies of Au-, CO-, and Xe-terminated tips interacting with 3,4,9,10-perylene-tetracarboxylic-dianhydride (PTCDA) molecules [9]. In this case, their use of the small-amplitude approximation for converting frequency shift to force (Eq. 6) is sufficiently justified. Weymouth et al. apply the small-amplitude approximation to data concerning Si-terminated tips interacting with adatoms of the Si(111)-(7 × 7) surface, using a qPlus sensor with $a$ = 400 pm [10]. Despite this amplitude being at least an order of magnitude too large for this approximation to be accurate with respect to the short-range chemical interaction (according to the calculations in Sect. 2.2), their model of 'phantom' forces agrees well with their experimental data, suggesting that the length scale of this 'phantom' force is much greater than the chemical interaction force.

An instance of the justified use of the large-amplitude approximation is given by Sawada et al. [11], where amplitudes of $a > 20$ nm have been used to probe adatoms of the Si(111)-(7 × 7) surface using a Si-terminated tip located at the free end of a soft ($k$ = 22.9 N/m) cantilever. In this case, the oscillation amplitude is more than an order of magnitude and greater than the characteristic length scale of the interaction potential. An instance when one must also take care that an oscillation amplitude is 'large' is when performing a commonly applied method of amplitude calibration [7, 12]. In the large-amplitude approximation, $\Delta f(z) \cdot a^{3/2}$ is solely dependent on $z$ (tip–sample separation at point of closest approach). As such, amplitude can be calibrated by relating piezo retraction to the required increase in amplitude to maintain $\Delta f(z) \cdot a^{3/2}$ constant. Our calculations highlight that this technique should only be applied to amplitudes an order of magnitude larger than the tip–sample interaction being probed for calibration. In this case, it is the total tip–sample interaction which is being probed, and one would expect the typical length scale of this to be greater than that of purely chemical interactions.

## 3 Numerical Evaluation of Formulae

To calculate an interaction force from frequency shift data, it is convenient to calculate the interaction potential first and then differentiate to obtain the corresponding force. One advantage to this procedure is that it separates the numerical integration and numerical differentiation stages, allowing greater flexibility in the choice of numerical evaluation technique. This approach is also sensible when dealing with three-dimensional 'grid spectroscopy' data sets, where the scalar potential field must be calculated first to permit further calculation of the corresponding vector force field (although this is beyond the scope of this chapter, we focus here solely on one-dimensional data sets).

To numerically evaluate an interaction potential from experimental frequency shift data, five pieces of information are required: the resonant frequency $f_0$, spring constant $k$, oscillation amplitude $a$, of the cantilever, the array of $N$ uniformly spaced tip–sample separations $\{z_1, z_2, \ldots, z_N\}$ (in order of increasing separation), and the corresponding array of $N$ measured frequency shifts $\{\Delta f_1, \Delta f_2, \ldots, \Delta f_N\}$. This data can then be used to generate the array of $N$ corresponding interaction potential values $\{U_1, U_2, \ldots, U_N\}$.

To calculate the interaction potential for an arbitrary oscillation amplitude, two aspects concerning the numerical evaluation of Eq. 4 need to be considered. The first concerns the improper integral [13]—the integrand diverges at its lower bound ($t = z$) due to the third term inside the brackets. As such, performing numerical integration techniques to this improper form will result in divergences. This problem can be resolved by changing the integration variable to $u = \sqrt{t - z}$, which gives

$$U(z) = \frac{4k}{f_0} \int_0^{\infty} \left( u^3 + \sqrt{\frac{a}{16\pi}} u^2 + \sqrt{\frac{a^3}{2}} \right) \Delta f(u^2 + z)\, \mathrm{d}u. \tag{11}$$

The integrand is now finite at the lower bound (the diverging integrand can also be dealt with via the use of correction terms [2]). The second issue arises due to the infinite upper bound of the integral—an experimental data set will not extend to an infinite tip–sample separation. This can be resolved by changing the upper bound to a tip–sample separation at which $\Delta f \approx 0$ (within experimental noise). At separations beyond this point, the integrand can be considered to be insignificantly small.

A consequence of applying the variable change discussed above is that the new integration variable has a non-uniform spacing. As such, numerical integration techniques of higher orders, e.g. Simpson's (2nd order) rule, cannot be applied. Therefore, using the trapezoidal rule with a non-uniform spacing, Eq. 11 can be numerically evaluated to give the potential $U_i$ at tip–sample separation $z_i$ as

$$U_i \approx \frac{4k}{f_0} \sum_{j=i}^{N-1}{}' \left( u_{i,j}^3 + \sqrt{\frac{a}{16\pi}} u_{i,j}^2 + \sqrt{\frac{a^3}{2}} \right) \Delta f_j \cdot \left( u_{i,j+1} - u_{i,j} \right), \tag{12}$$

where $u_{i,j} = \sqrt{z_j - z_i}$, the prime ($'$) indicates half-weighting of the $j = i$ term, and $U_N \approx 0$.

Once an array of potential values has been obtained, the corresponding array of force values may be calculated using the original array of tip–sample separations. Whilst standard numerical integration techniques are fairly robust with respect to noisy data, standard numerical differentiation techniques will amplify noise present in data. One approach to counter such effects is to use a low-noise Lanczos differentiator, which in this case may be expressed as

$$F_i \approx -\frac{3}{\Delta z} \sum_{j=1}^{m} \frac{j\left(U_{i+j} - U_{i-j}\right)}{m(m+1)(2m+1)} \tag{13}$$

where $\Delta z$ is the (constant) spacing between successive tip–sample separations, and $m = 1, 2, 3, \ldots$ relates to the region size over which data are included in the differentiator (with $m = 1$, Eq. 13 is identical to the lowest order central differences differentiator—a poor-noise suppressing method). Increasing $m$ increases the level of noise suppression; however, one must ensure that $m\Delta z$ is less than any characteristic length scale of the potential or the calculated force may be distorted (e.g. the maximum attractive force may be underestimated).

## 4 Isolating Short-Range Interactions

There are many different tip–sample interactions that may be significant in any given DFM experiment. For convenience, here, we simply categorize these interactions as either short-range or long-range. We term short-range interactions as those that are site specific (dependent on lateral position) and depend on the chemical nature of the tip apex and surface directly beneath. We term long-range (background) interactions as those that are non-site specific (independent of lateral position) and depend on the bulk composition and geometry of tip and sample (e.g. van der Waals interactions). In this scheme, the total tip–sample interaction force,

$F_T$, is simply the sum of short-range, $F_{SR}$, and long-range, $F_{LR}$, interaction forces, $F_T = F_{SR} + F_{LR}$. The analysis procedures presented here concern the extraction of this short-range component of tip–sample interaction forces.

In DFM, force is not a directly measured property, but obtained via an integration of frequency shift (see Sect. 2). The total tip–sample frequency shift, $\Delta f_T$, can be thought of as the sum of frequency shift components due to short-range, $\Delta f_{SR}$, and long-range, $\Delta f_{LR}$, interactions, $\Delta f_T = \Delta f_{SR} + \Delta f_{LR}$. To isolate short-range interaction forces from DFM spectroscopy measurements, two processes must be accomplished: subtraction of long-range contributions and conversion from frequency shift to force. These steps can occur in either order:

- $\Delta f_{LR}$ subtracted from $\Delta f_T$ gives $\Delta f_{SR}$, which is integrated to give $F_{SR}$.
- $\Delta f_T$ is integrated to give $F_T$, from which $F_{LR}$ is subtracted to give $F_{SR}$.

Formulae for converting frequency shift to force can only be applied to data which extend to tip–sample separations large enough such that $\Delta f \approx 0$. This is not an issue if $\Delta f_{SR}$ is used, which will become insignificantly small at tip–sample separations typically not greater than one nanometre from the sample. $\Delta f_T$ will typically only become insignificantly small after many (even tens of) nanometres due to long-range contributions. As such, to calculate the total tip–sample interaction force, frequency shift measurements need to extend to large tip–sample separations. Therefore, it is more practical to perform the long-range component subtraction in the frequency shift domain and then integrate to obtain the short-range interaction force (as opposed to integrating total frequency shift data to give total interaction force and then subtract the long-range contribution in the force domain).

## *4.1 Methods of Background Subtraction*

There are two common methods of subtracting frequency shift contributions due to long-range interactions—referred to here as the 'extrapolation' method and the 'on–off' method. The extrapolation method involves acquiring a frequency shift spectroscopy measurement over the feature of interest out to large tip–sample separations. Via inspection of this data (frequency shift against tip–sample separation), one estimates the separation at which short-range interactions become insignificant —a point that beyond which the measurement is solely sensitive to long-range interactions. This long-range data is fit, typically to a power law with at least three free parameters, and extrapolated towards the surface to estimate the long-range contribution to the total tip–sample interaction at separations where short- and long-range contributions are inseparable. This extrapolated fit is then subtracted from the original data to estimate the frequency shift solely due to short-range interactions. There are several reasons why this technique has the potential to introduce significant uncertainties. For example, extrapolating a rapidly diverging power law will dramatically magnify any slight error in fitting. Furthermore, it is naïve to expect that any given frequency shift data should conform to simple fitting, a 'good' fit to long-range data may simply be due to the fitting flexibility of power laws with free parameters. A quantitative analysis of the uncertainties of this method is beyond the scope of this work and is presented elsewhere [14, 15].

The on–off method involves acquiring two frequency shift spectroscopy measurements—one on, and one off, the feature of interest, encompassing the same range of tip–sample separations (only out to approximately 1–2 nm is typically sufficient). The 'on' measurement will contain information on short- and long-range interactions ($\Delta f_{\mathrm{on}} = \Delta f_{\mathrm{SR}} + \Delta f_{\mathrm{LR}}$), and ideally, the 'off' measurement will only contain information on long-range interactions ($\Delta f_{\mathrm{off}} = \Delta f_{\mathrm{LR}}$), resulting in the difference between these measurements being solely due to short-range interactions, $\Delta f_{\mathrm{SR}} = \Delta f_{\mathrm{on}} - \Delta f_{\mathrm{off}}$, due to the non-site specific nature of long-range interactions [16, 17]. Here, we concern ourselves with experimental instances where it is possible to obtain 'off' measurements which solely contain long-range contributions. One such instance is probing adatoms of the Si(111)-(7 × 7) surface, where the 'on' measurement is acquired over an adatom and the 'off' measurement is acquired over a cornerhole (a surface feature with an absence of adatoms). Another instance is for protruding surface-adsorbed molecules—one measurement can be acquired over a position on a molecule and the other measurement can be acquired off to the side of the molecule. Examples of these two scenarios can be found in Chiutu et al. [18]. For certain samples, it is not possible to obtain an 'off' measurement with only long-range contributions. For example, the Si(100) surface has no suitable null points. In situations like this, the on–off method is limited to calculating the difference in the short-range interactions between two different sites (see, for example Kichin et al. [9] and Sharp et al. [19] for instances when force differences are calculated).

## 4.2 Processing Experimental Data

The on–off method is not completely devoid of potential errors. Typically, two frequency shift measurements will not initially be aligned in the $z$-axis, due, for example, to different feedback heights over the two measurement positions [20]. As such, the $z$-axis of one measurement needs to be shifted with respect to the other. In this section, we discuss this aligning process applied to experimental data. Here, we present a transparent perspective on the complete process of extracting a short-range interaction force from DFM measurements.

The following data were acquired using an Omicron Nanotechnology GmbH qPlus DFM/STM operating in an LHe/$LN_2$ bath cryostat with a sample temperature of approximately 77 K due to $LN_2$ cooling. A commercial qPlus sensor (tungsten wire attached to the end of a quartz tuning fork) was used, with resonant frequency $f_0 = 24{,}372.2\,\mathrm{Hz}$, spring constant $k \approx 2000$ N/m (estimated [21]), and oscillation amplitude $a \simeq 275\,\mathrm{pm}$. The sample investigated here is the Si(111)-(7 × 7) surface, freshly prepared following standard flashing/annealing procedures. The tip was prepared via standard pulsing/crashing methods, and thus, it is expected to be terminated by silicon (and not tungsten). Thermal drift was minimized using an atom-tracking system [22]. Piezoelectric creep was minimized by scanning the same region for several hours prior to measurements and by acquiring data during tip retractions (which preceded a small tip approach to the closest tip-sample separation). The tip–sample gap bias was held at 0 V to eliminate any 'phantom' forces [10].

Figure 4a shows the raw data of two spectroscopy measurements, one obtained over an Si adatom and the other over a cornerhole. In both measurements, $z = 0$ represents the imaging feedback height over that position (the surface was imaged using constant frequency shift feedback, with $\Delta f = -50.2\,\text{Hz}$ set point). Since the two measurement positions have different tip–sample interactions, and thus different feedback heights, these two measurements are not correctly aligned relative to one another. Thus, one must determine the correct tip–sample separation shift, $z_{\text{shift}}$, to align these data. Here, we adjust the $z$-axes of 'on' measurements and maintain the $z$-axes of 'off' measurements (the alternative is possible, but less convenient, since multiple 'on' measurements can use the same 'off' measurement as a reference).

The calculation of $z_{\text{shift}}$ is accomplished by aligning the long-range component of the 'on' measurement with the 'off' measurement (since the latter is assumed to be entirely due to long-range interactions). However, by simple inspection of the 'on' spectroscopy measurement, it is not clear where the dividing cut-off, $z_{\text{cut-off}}$, lies (above which only long-range interactions contribute, i.e. $\Delta f_{\text{on}} = \Delta f_{\text{LR}}$ for $z > z_{\text{cut-off}}$). As such, one should trial a range of $z_{\text{cut-off}}$ values to make a confident decision of the correct value of $z_{\text{shift}}$.

For a given trial value of $z_{\text{cut-off}}$, $z_{\text{shift}}$ is calculated in the following way. The 'on' measurement data above $z_{\text{cut-off}}$ are compared to the 'off' measurement data over the tip–sample separation values which are common to both data sets. The 'on' data can be shifted by integer amounts of $\Delta z$, the interval spacing of data points on the $z$-axis. $z_{\text{shift}}$ is then given by the amount of shifting which minimizes the sum of squared differences between the 'on' and 'off' data—for the correct value of $z_{\text{shift}}$, the residual should simply fluctuate around zero over the entire long-range segment of tip–sample separations. This alignment technique is shrewd due to its interpolation-free nature, and uncertainty in alignment is minimized to $\delta z_{\text{shift}} \approx \Delta z$.

Figure 4b shows a plot of $z_{\text{shift}}$ values obtained for a range of $z_{\text{cut-off}}$ values—three regions can be identified. When $z_{\text{cut-off}}$ is too large, there is little overlap between the shifted 'on' and 'off' measurements, resulting in noisy variation of $z_{\text{shift}}$. When $z_{\text{cut-off}}$ is too small, short-range data are included in the truncated 'on' measurement and the resulting alignment is flawed. At intermediate values of $z_{\text{cut-off}}$, there are sufficient data points for a comparison without the introduction of short-range interactions. Over this range, $z_{\text{shift}}$ is insensitive to the precise selection of $z_{\text{cut-off}}$, and it is this value ($z_{\text{shift}} = 34.1$ pm here, compared to $31 \pm 5$ pm as estimated from the difference in feedback heights of the surface scan) which is determined as the correct amount to shift the 'on' measurement by to align with the 'off' measurement.

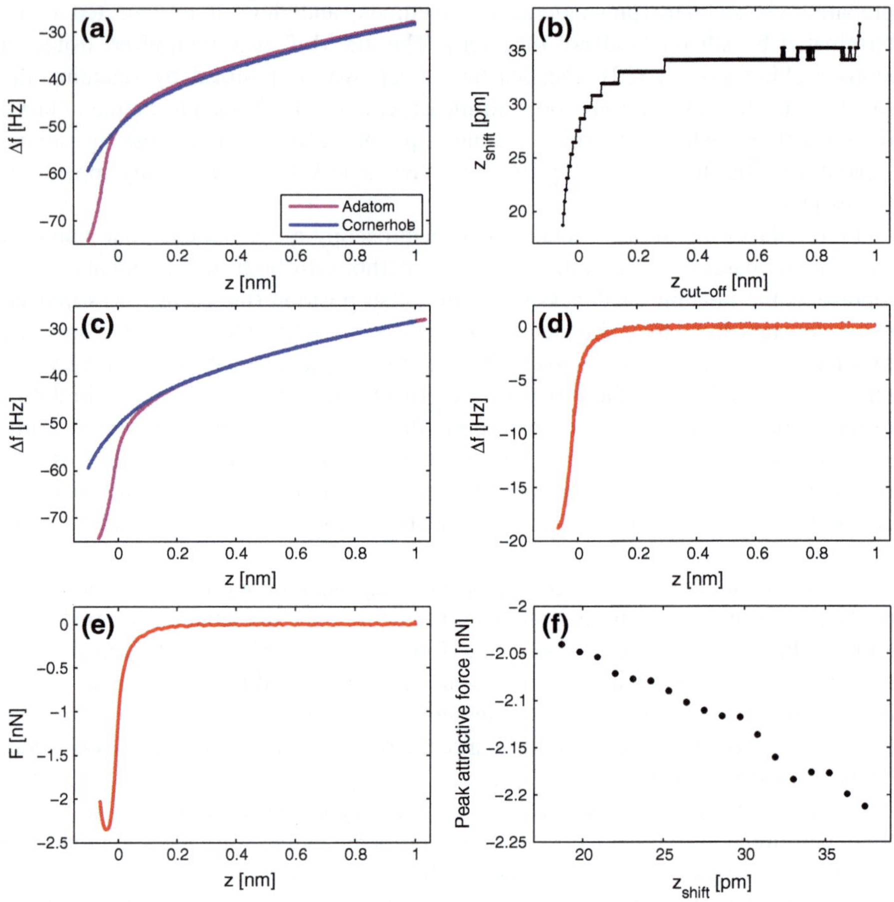


**Fig. 4** A transparent approach to background subtraction to isolate the short-range interaction force between an Si-terminated tip and an adatom of the Si(111)-(7 × 7) surface. **a** Raw spectroscopy data measured over an adatom and a cornerhole. **b** Variation in $z_{shift}$ due to selection of $z_{cut-off}$. **c** Spectroscopy measurements aligned by applying $z_{shift} = 34.1$ pm to the 'on' measurement. **d** Short-range frequency shift data obtained by subtraction the 'off' measurement from the aligned 'on' measurement. **e** Calculated short-range interaction force, with $m = 10$ used in the Lanczos differentiator. **f** Variation in the peak attractive force depending on the value of $z_{shift}$

Figure 4c shows the spectroscopy data aligned using this informed choice of $z_{shift}$, with Fig. 4d showing the short-range frequency shift component obtained by subtracting the 'off' measurement from the aligned 'on' measurement. Another indicator that the alignment appears reliable is that the tail data are flat and fluctuate around zero. Using the numerical technique described in Sect. 3, the short-range interaction force can be calculated, which is shown in Fig. 4e and can be seen to be in very good agreement with previous experimental and theoretical studies [5, 16, 23]. Finally, it is informative to see the result of using different $z_{\text{shift}}$ values on the extracted interaction force. Figure 4f shows how the maximum attractive force changes when different $z_{\text{shift}}$ values are used. One can see there is variation of around 20 pN/pm—not an insignificant amount. The data shown here have a minimum uncertainty in $z_{\text{shift}}$ of $\delta z_{\text{shift}} \approx 1$ pm, and as such, there is a minimum uncertainty in force due to the 'on–off' of approximately 20 pN (a small, 1 %, uncertainty).

## 5 Summary

The use of DFM to perform high-accuracy measurements of short-range chemical interactions between well-defined tip- and surface-adsorbed species is becoming increasingly commonplace. For these types of experiments to be quantitatively robust, a critical eye must be cast over the sources of uncertainty in any given measurement. An obvious source of error is due to miscalibration. However, even for a well-calibrated system, the analysis procedure used to convert experimentally acquired total frequency data to short-range tip–sample interaction forces must be undertaken with the utmost care and attention. Here, we have detailed a considered approach of how to perform the subtraction of long-range contributions and the conversion from frequency shift to force.

With respect to the subtraction of long-range contributions, a rigorous implementation of the 'on–off' method was described. Using this technique, uncertainties can arise due to the alignment of $z$-axes of the 'on' and 'off' measurements and the determination of the tip–sample separation cut-off (above which only long-range interactions are significant). We described a rigorous approach to find this cut-off separation and perform the optimum $z$-axes alignment and in doing so were able to quantify the (very small) uncertainties arising from implementation of this subtraction technique. In regard to the conversion from the frequency shift to the force domain, we explored the validity of approximation formulae relating to different regimes of oscillation amplitude, concluding that the small[large]-amplitude approximation should only be made when the amplitude is at least an order of magnitude less than [greater than] any characteristic length scale of the interaction being probed. Also detailed were numerical methods of integration and differentiation that could be used to tackle the formulae of DFM.

**Acknowledgments** The authors acknowledge Philip Moriarty, Peter Sharp, Samuel Jarvis, and Philipp Rahe for fruitful discussions. A.S. acknowledges financial support from the University of Nottingham through the award of a Nottingham Advance Research Fellowship. A.M.S. acknowledges financial support from the Leverhulme Trust through fellowship F00/114 BI.

# References


1. Giessibl, F.J.: A direct method to calculate tip-sample forces from frequency shifts in frequency-modulation atomic force microscopy. Appl. Phys. Lett. **78**, 123–125 (1998)
2. Welker, J., Illek, E., Giessibl, F.J.: Analysis of force-deconvolution methods in frequency-modulation atomic force microscopy. Beilstein J. Nanotechnol. **3**, 238–248 (2012)
3. Sader, J.E., Jarvis, S.P.: Accurate formulas for the interaction force and energy in frequency modulation force spectroscopy. Appl. Phys. Lett. **84**, 1801–1803 (2004)
4. Giessibl, F.J.: High-speed force sensor for force microscopy and profilometry utilizing a quartz tuning fork. Appl. Phys. Lett. **73**, 3956–3958 (1998)
5. Pérez, R., Štich, I., Payne, M.C., Terakura, K.: Surface-tip interactions in noncontact atomic-force microscopy on reactive surfaces: Si(111). Phys. Rev. B **58**, 10835–10849 (1998)
6. Albrecht, T.R., Grütter, P., Horne, D., Rugar, D.: Frequency modulation detection using high-Q for enhanced force microscope sensitivity. J. Appl. Phys. **69**, 668–673 (1991)
7. Simon, G.H., Heyde, M., Rust, H.-P.: Recipes for parameter determination in dynamic force spectroscopy: spring constant and amplitude. Nanotechnology **18**, 255503 (2007)
8. Dürig, U.: Relation between interaction force and frequency shift in large-amplitude dynamic force microscopy. Appl. Phys. Lett. **75**, 433–435 (1999)
9. Kichin, G., Wagner, C., Tautz, F.S., Temirov, R.: Calibrating atomic-scale force sensors installed at the tip apex of a scanning tunneling microscope. Phys. Rev. B **87**, 081408(R) (2013)
10. Weymouth, A.J., Wutscher, T., Welker, J., Hofmann, T., Giessibl, F.J.: Phantom force induced by tunneling current: a characterization on Si(111). Phys. Rev. Lett. **106**, 226801 (2011)
11. Sawada, D., Sugimoto, Y., Morita, K., Abe, M., Morita, S.: Simultaneous measurement of force and tunneling current at room temperature. Appl. Phys. Lett. **94**, 173117 (2009)
12. Giessibl, F.J., Bielefeldt, H.: Physical interpretation of frequency-modulation atomic force microscopy. Phys. Rev. B **61**, 9968–9971 (2000)
13. Press, W.H., Teukolsky, S.A., Vetterling, W.T., Flannery, B.P.: Numerical Recipes: The Art of Scientific Computing, 3rd edn, pp. 167–172. Cambridge University Press, Hong Kong (2007)
14. Sweetman, A.M., Stannard, A.: Uncertainties in forces extracted from non-contact atomic force microscopy measurements by fitting of long-range background forces. Beilstein J. Nanotechnol. **5**, 386–393 (2014)
15. Kuhn, S., Rahe, P.: Discriminating short-range from van der Waals forces using total force data in noncontact atomic force microscopy. Phys. Rev. B **89**, 235417 (2014)
16. Lantz, M.A., Hug, H.J., Hoffmann, R., van Schendel, P.J.A., Kappenberger, P., Martin, S., Baratoff, A., Güntherodt, H.-J.: Quantitative measurement of short-range chemical bonding forces. Science **291**, 2580–2583 (2001)
17. Ternes, M., González, C., Lutz, C.P., Hapala, P., Giessibl, F.J., Jelínek, P., Heinrich, A.J.: Interplay of, force, and structural change in metallic point contacts. Phys. Rev. Lett. **106**, 016802 (2011)
18. Chiutu, C., Sweetman, A.M., Lakin, A.J., Stannard, A., Jarvis, S., Kantorovich, L., Dunn, J.L., Moriarty, P.: Precise orientation of a single $C_{60}$ molecule on the tip of a scanning probe microscope. Phys. Rev. Lett. **108**, 268302 (2012)
19. Sharp, P., Jarvis, S., Woolley, R., Sweetman, A., Kantorovich, L., Pakes, C., Moriarty, P.: Identifying passivated dynamic force microscopy tips on H:Si(100). Appl. Phys. Lett. **100**, 233120 (2012)
20. Sugimoto, Y., Pou, P., Custance, Ó., Jelinek, P., Morita, S., Pérez, R., Abe, M.: Real topography, atomic relaxations, and short-range chemical interactions in: the case of the $\alpha$-Sn/Si(111)-($\sqrt{3} \times \sqrt{3}$)R30° surface. Phys. Rev. B **73**, 205329 (2006)
21. Sweetman, A., Jarvis, S., Danza, R., Bamidele, J., Gangopadhyay, S., Shaw, G.A., Kantorovich, L., Moriarty, P.: Toggling bistable atoms via mechanical switching of bond angle. Phys. Rev. Lett. **106**, 136101 (2011)

22. Rahe, P., Schütte, J., Schniederberend, W., Reichling, M., Abe, M., Sugimoto, Y., Kühnle, A.: Flexible drift-compensation system for precise 3D force mapping in severe drift environments. Rev. Sci. Instrum. **82**, 063704 (2011)
23. Yurtsever, A., Sugimoto, Y., Tanaka, H., Abe, M., Morita, S., Ondráček, M., Pou, P., Pérez, R., Jelínek, P.: Force mapping on a partially H-covered Si(111)-(7 × 7) surface: influence of tip and surface reactivity. Phys. Rev. B **87**, 155403 (2013)

21. Rahe, P., Schütte, J., Schniederberend, W., Reichling, M., Abe, M., Sugimoto, Y., Kühnle, A.: Flexible drift-compensation system for precise 3D force mapping in severe drift environments. Rev. Sci. Instrum. 82, 063704 (2011)
22. Yurtsever, A., Sugimoto, Y., Tanaka, H., Abe, M., Morita, S., Ondráček, M., Pou, P., Pérez, R., Jelínek, P.: Force mapping on a partially H-covered Si(111)-(7 × 7) surface: Influence of tip and surface reactivity. Phys. Rev. B 87, 155403 (2013)

# Theoretical Challenges of Simultaneous nc-AFM/STM Experiments

**P. Jelínek**

**Abstract** The invention of scanning probe microscopy (SPM) provided an unprecedented method to explore locally the material properties of surfaces and nanostructures at the atomic scale. Recent progress, involving the simultaneous acquisition of tunneling current and atomic force measurements, opens new horizons in the SPM field. On the other hand, this enhanced experimental complexity calls for a detailed understanding of the imaging mechanisms. Reliable interpretation of experimental data and further proliferation of this technique strongly relies on our theoretical understanding of ongoing physical processes during scanning. Here, we discuss the current status of, and challenges for, the theoretical description of simultaneous AFM/STM measurements.

**Keywords** Atomic force microscopy · Scanning tunneling microsopy · Density functional theory · Atomic contrast

## 1 Introduction

Recent progress in scanning probe microscopy has allowed simultaneous scanning tunneling microscopy (STM) [1] and atomic force microscopy (AFM) [2] measurements to be acquired with atomic resolution. This combination brings completely new possibilities not only for more complex characterization of surfaces and nanostructures, but it also opens new horizons for understanding fundamental processes such as formation of chemical bonds [3] or charge transfer [4] at atomic scale.

STM is a well-established technique, routinely used to image, characterize, and modify objects at the atomic scale [5]. Here, the tunneling current flowing between an atomically sharp probe and a surface is employed as the experimental observable. Consequently, application of STM is limited to only conductive samples.

P. Jelínek (✉)
Institute of Physics, Academy of Sciences of the Czech Republic,
Cukrovarnická 10, 16200 Prague, Czech Republic
e-mail: jelinekp@fzu.cz

 
P. Moriarty and S. Gauthier (eds.), *Imaging and Manipulation of Adsorbates Using Dynamic Force Microscopy*, Advances in Atom and Single Molecule Machines,
DOI 10.1007/978-3-319-17401-3_5

This serious limitation can be overcome using AFM, where the detected signal is the force acting between the outermost atoms of the tip and the sample. The possibility of exploring non-conductive samples with high resolution at the nanoscale led to AFM being adopted as an important tool in many areas (e.g., biology, chemistry, and physics) in the last decade. In particular, the so-called frequency modulation AFM (FM-AFM) [6] technique became widely used under ultrahigh vacuum (UHV) conditions due to its high sensitivity in detecting forces at the atomic scale. Here, the modulation of the frequency of the oscillating probe, proportional to the force gradient acting between the tip and the sample, is recorded, while the oscillation amplitude is kept constant. The technique has shown great ability to provide high-resolution images [7], chemical sensitivity [8], or single atom manipulation on all types of surfaces (see, for example, chapter "Atom Manipulation Using Atomic Force Microscopy at Room Temperature" of this volume by Sugimoto), including insulators [9].

Recently, simultaneous FM-AFM/STM measurements of both atomic force and tunneling current with atomic resolution using different sensors and systems (many commercially available) have been reported (see, e.g., [10–13]). It is evident that we have reached the stage at which precise simultaneous measurements of the force and the tunnel current are "de rigueur" in the scanning probe community.

This combination of STM and AFM is, of course, not without its challenges and technical difficulties. STM tends to operate in a tip–sample distance regime where perturbation of the atomic structure due to the proximity of the probe is negligible (see Fig. 1). This is not the case in AFM, where the probe used to operate in the

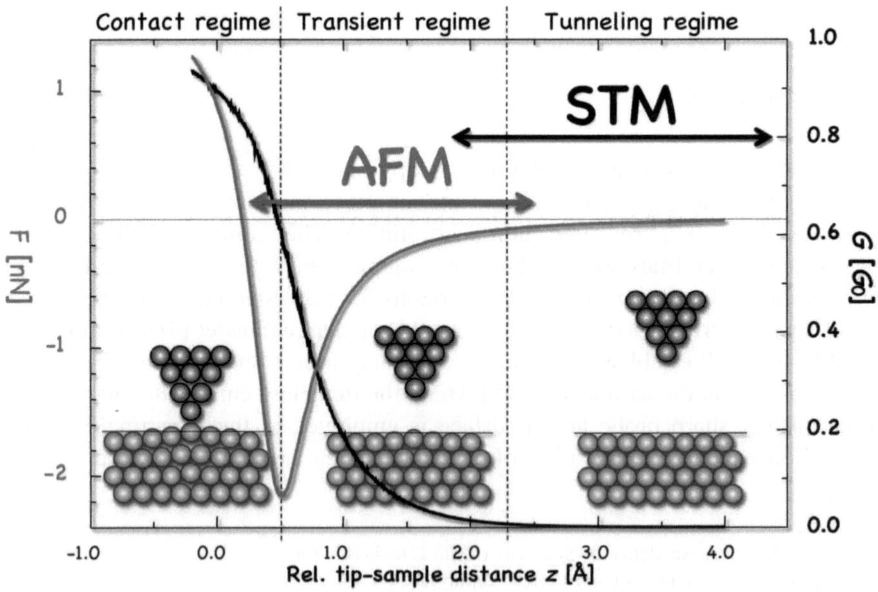


**Fig. 1** Schematic illustration representing three different tip–sample interaction regimes depending on distance

attractive regime, and the formation of a chemical bond between the atoms at the apex of the tip and on the sample surface is often responsible for the observed atomic contrast. Therefore, atomic relaxation cannot be omitted in the analysis of simultaneous AFM/STM experiments. In other words, applications of frequently adopted theoretical approaches to simulate STM images, such as Tersoff–Hamann [14] or Bardeen [15, 16] method, are questionable here. On the other hand, the presence of the tunneling current may introduce an extra electrostatic force [17], which is not present in standard AFM measurements. What is more, simultaneous acquisition of the tunneling current and the frequency shift might introduce undesired cross talk between these two channels. Therefore, special attention must be paid to understand and control this phenomenon [18].

The aim of this chapter was to provide a detailed theoretical description of the processes taking place during simultaneous AFM/STM measurements with atomic resolution. In particular, we will pay special attention to the formation of chemical bonds between the outermost atoms of the tip and the sample and its consequences on the behavior of the tunneling current and other quantities.

## 2 Theoretical Description of Scanning Probe Microscopy

### 2.1 Atomic Force Microscopy

Typically, theoretical AFM simulations involve calculation of the short-range forces acting between the tip and the sample as a function of their distance [19]. The resulting force–distance curves calculated over different characteristic surface sites can be compared directly or indirectly to the experimental evidence (see, e.g., [20–25]). The character of the short-range force is determined by the chemical composition of both the tip and the sample. Generally, on metallic or semiconductor surfaces, the short-range forces reflect the chemical bond established between the outermost atoms of the tip and the sample [26]. Therefore, only simulations employing quantum mechanics can properly describe the mechanism of formation of the chemical bond. On the other hand, precise quantum mechanical calculations are computationally very demanding, which strongly limits the size and the complexity of the modeled systems. From this point of view, simulations based on density functional theory (DFT) seem to be the optimal choice as DFT provides the optimal ratio between the accuracy and the size of model systems.

Typically, we carry out a series of DFT calculations, incorporating several hundreds of atoms, which simulating a tip approach toward the surface. Here, a small cluster of atoms represents a probe and a supercell model including slabs and vacuum is the more efficient choice to simulate the solid surface (see Fig. 2). The slab in the supercell is infinite and periodic in the directions parallel to the surface, but finite in the direction perpendicular to the surface. Subsequently, the tip model is statically shifted toward the surface slab, and selected atoms are relaxed to their

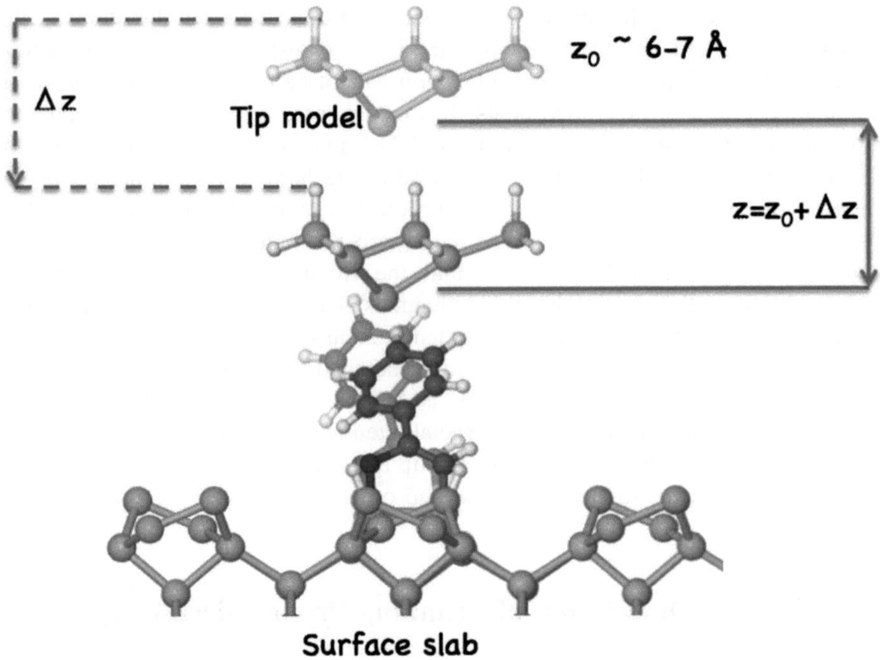


**Fig. 2** Schematic illustration of an atomistic model of a tip and sample for AFM simulations. The tip, represented by a small cluster of atoms, is statically approached toward the surface slab by gradually decreasing the tip–sample distance z from its initial value $z_0$ by a constant decrement, Δz

optimal positions (see Fig. 2) until the residual forces acting on the atoms are smaller than a threshold value, typically less than 0.05 eV/Å.

As already mentioned above, the resulting short-range forces strongly rely on the chemical composition of the tip. For example, Fig. 3 shows the short-range forces calculated over a silicon adatom on the Si(111)-7 × 7 surface using three different tip models. More details about the computational procedure and a comparison to experimental evidence can be found in Yurtsever et al. [27]. The three tip models consist of a silicon cluster [28] with different apex terminations, and they can be viewed as representative of the different interaction regimes available on semiconductor surface:

1. *Strong covalent bond regime*
   A bare silicon tip (black curve in Fig. 3a) gives the larger attractive force. Pronounced accumulation of electron density between the atoms at the very apex of the tip and those at the surface indicates the presence of a strong covalent bond. This conclusion is also supported by a strong modification of the local density of states (LDOS), as shown in Fig. 3b (black).
2. *Short-range electrostatic force regime*

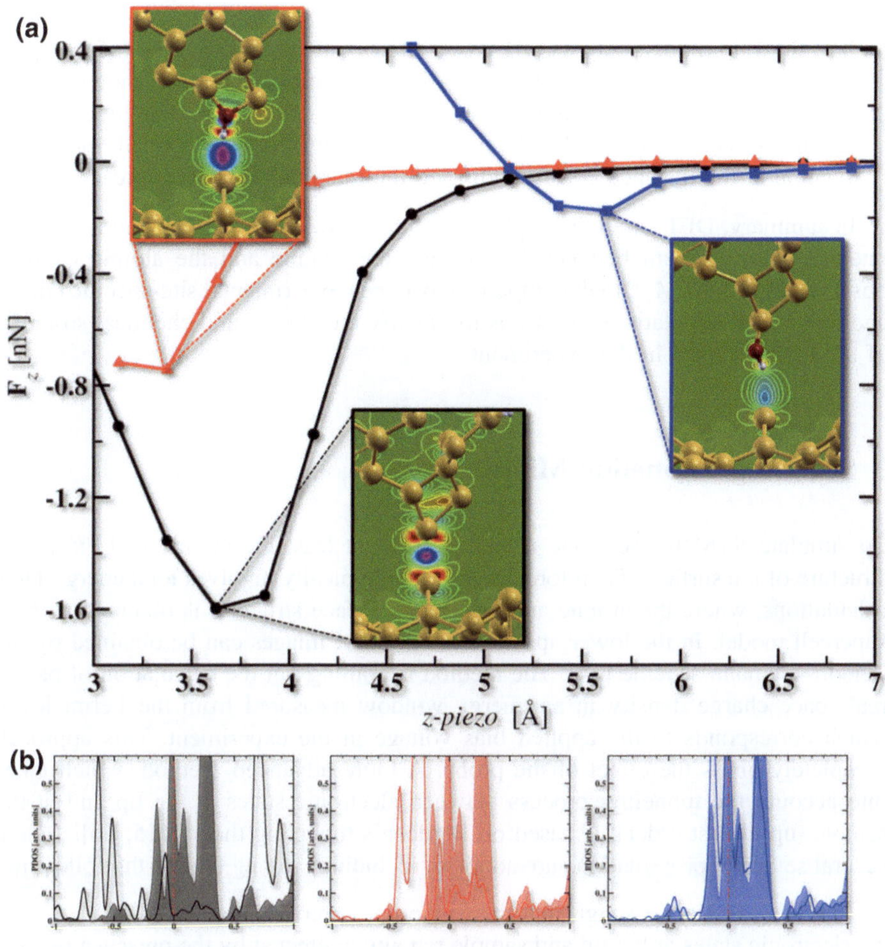


**Fig. 3** **a** Three different force curves calculated over a silicon adatom on the Si(111)-7 × 7 surface using different tip models (see *inset* figures). The *inset* figures show induced electron densities between tip and surface atoms corresponding to distances where the maximal attractive force occurs. **b** Plot the corresponding variation of the LDOS at the maximal attractive force (*empty*) with respect to the unperturbed states (*filled*) for the three different cases

In the case of the Si-tip model with an OH group, where the oxygen atom is connected to two Si atoms underneath, we observed significantly smaller attractive forces (see red curve in Fig. 3a) and less pronounced electron density accumulation between the tip and the sample. The LDOS of the Si adatom shown in Fig. 3b (red color) reveals only modest modification, mainly consisting of a rigid movement in energy. This observation indicates the dominance of an electrostatic interaction instead of the presence of a true covalent bond.

3. *Non-reactive regime*
   The third tip model, with an OH group attached to a single Si atom of the apex, represents a very weakly interacting tip. The interaction force shows a very weak bonding minimum, and it enters quickly into the repulsive regime at close distances. The calculated LDOS of the Si adatom reveals practically no modification at the distance corresponding to the maximal attractive force.

In summary, DFT simulations play a significant role in our understanding of the interaction mechanism between the tip and the sample and the atomic contrast observed in FM-AFM. Direct comparison between experimental site-specific curves and theoretical simulations allows us to identify the atomic and chemical structure of the tip apex used in the experiment.

# 3 Scanning Tunneling Microscopy

To simulate STM images, one needs to know at least the atomic and electronic structure of the surface. Therefore, the first step typically involves total energy DFT calculations, where the atomic and electronic surface structure is obtained using a supercell model. In the lower approximation, STM images can be obtained by the Tersoff–Hamann scheme [14]. The method is nothing but the calculation of partial real space charge density in an energy window measured from the Fermi level, which corresponds to the applied bias voltage in the experiment. This approach completely omits the effect of the probe. A more advanced method, which takes into account the tunneling process between electronic states at the tip and at the sample (up to first order), is based on Bardeen's tunneling theory [15, 16]. Again, several severe approximations are adopted, including, among others, the following:

- tunneling is weak enough that the first-order approximation is valid;
- electronic states at the tip and sample remain unaffected by the presence of each other, i.e., they are orthogonal to each other;
- atomic structures of the tip and sample remain unperturbed;
- the electron–electron interaction in the tunneling junction is ignored;
- electron occupations of the tip and sample are independent and constant;
- the tip and the sample are each in electrochemical equilibrium.

Here, the tunneling matrix elements are related to the derivatives of the sample wavefunctions in a simple and straightforward way in terms of localized orbitals [29].

To take into account out-of-equilibrium conditions and multiple electron scattering events, a Green's function (GF) technique was devised [30]. For example, it has been shown that multiple electron scattering may significantly change the atomic resolution of STM images at close distances [31]. For a more detailed discussion about the effect of multiple electron scattering on electron transport between the frontiers orbital of the tip and the sample see, for example, Ref. Jelínek et al. [32].

Finally, we should stress that a precise modeling of simultaneous AFM/STM experiments is only possible if we adopt the GF method on the fully relaxed atomic structure of both the tip and the sample, which takes into account the atomic relaxation due to the forces acting between the tip and the sample [14, 33].

# 4 Formation of Atomic Contact Between the Tip and the Sample

Let us discuss now what exactly happens when a chemical bond is formed between the tip apex and surface atoms. We should stress that the scenario described below is general for reactive tips on both semiconductor and metal surfaces. Figure 4 compares the calculated tunneling barrier, the short-range force, the conductance, and the induced electronic density as a metal tip approaches a Cu adatom on the Cu(111)

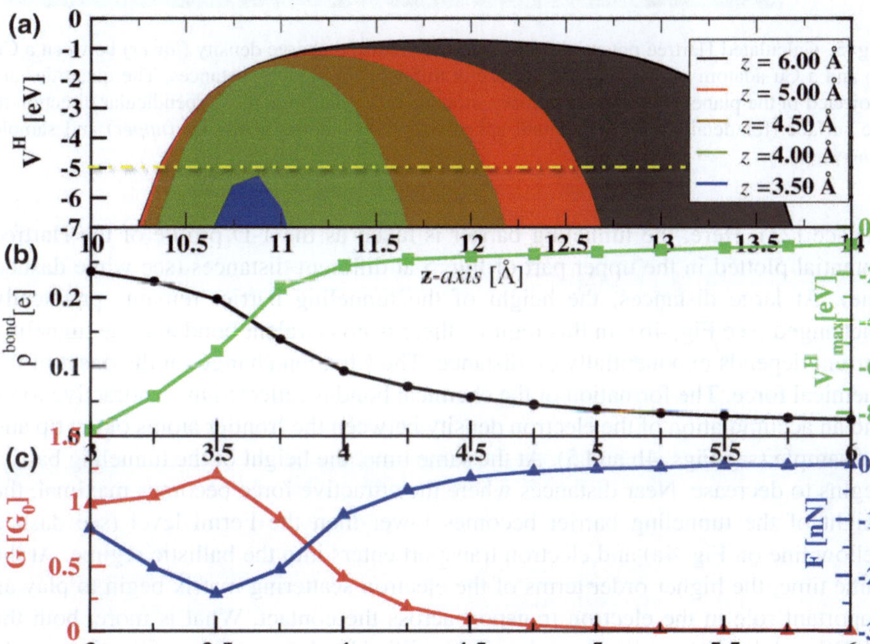


**Fig. 4** **a** Profile of the tunneling barrier between a Cu tip and a Cu adatom on the Cu(111) surface at different tip–sample distances (see *inset* figure); **b** evolution of the bond charge (*black*) and the tunneling barrier height (*green*) as function of tip–sample distance (x-axis); **c** the corresponding short-range force (*blue*) and tunneling current (*red*). More details can be found in Ternes et al. [23]

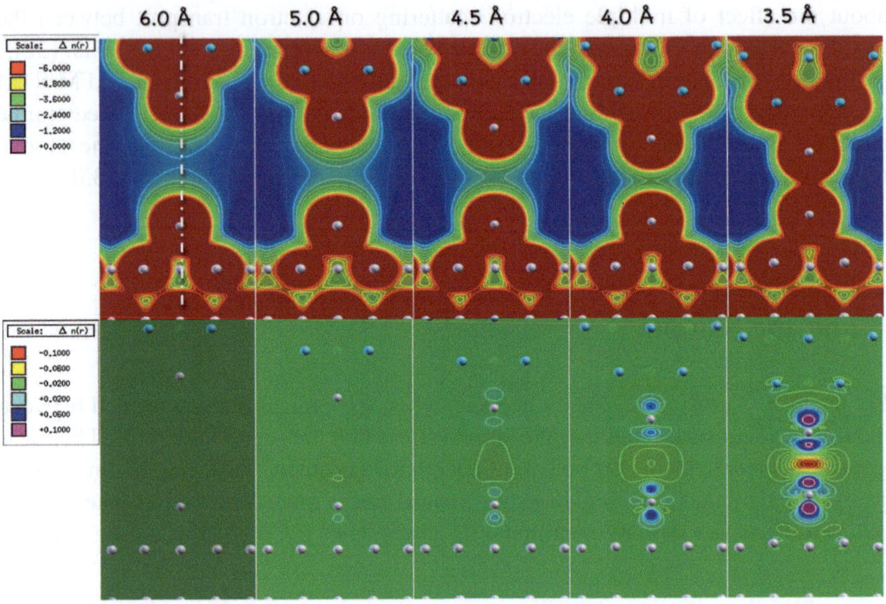


**Fig. 5** Calculated Hartree potential (*upper*) and the induced charge density (*lower*) between a Cu tip and a Cu adatom on the Cu(111) surface at different tip–sample distances. The quantities are projected in the plane including the tip apex atom and the adatom in the perpendicular direction to the surface (for details see [23]). Small spheres represent atoms of the tip (*upper*) and sample (*lower*)

surface [23]. Here, the tunneling barrier is taken as the 1-D profile of the Hartree potential plotted in the upper part of Fig. 5 at different distances (see white dashed line). At large distances, the height of the tunneling barrier remains practically unchanged (see Fig. 4b). In this regime, there is no covalent bond and the tunneling current depends exponentially on distance. The situation changes at the onset of the chemical force. The formation of the chemical bond is reflected in an attractive force and an accumulation of the electron density between the frontier atoms of the tip and the sample (see Figs. 4b and 5). At the same time, the height of the tunneling barrier begins to decrease. Near distances where the attractive force becomes maximal, the height of the tunneling barrier becomes lower than the Fermi level (see dashed yellow line on Fig. 4a) and electron transport enters into the ballistic regime. At the same time, the higher order terms of the electron scattering matrix begin to play an important role in the electron transport across the contact. What is more, both the LDOS and the surface dipole are locally modified by the proximity of the tip apex. In particular, the local modification of the surface states on semiconductor surfaces can lead to a vanishing of the tunneling current, as was shown on the Si(111)-7 × 7 surface both experimentally [12] and theoretically [12, 34]. Consequently, the conductance no longer depends on the tip–sample separation. It has also been shown that the local variation of the surface dipole due to the formation of a covalent bond is responsible for the atomic resolution observed by Kelvin probe force microscopy on semiconductor and ionic surfaces [35].

To summarize, we observe that the formation of a chemical bond is accompanied by the following events:

- pronounced atomic relaxation of both surface and tip atoms;
- modification of the tunneling barrier height, which completely vanishes at the maximal attractive force;
- saturation of the tunneling current and non-exponential behavior;
- local modification of the surface dipole due to a redistribution of the electron density at the atomic contact;
- local modification of the electronic states at both the tip and the sample.

One can therefore query the reliability of theoretical simulations of simultaneous nc-AFM/STM on atomically sized contacts on metallic surfaces. To answer this question, we can compare the calculated values of the conductance and the short-range force with the experimental evidence (for more details, see Ref. Ternes et al. [23]). As can be seen in Fig. 6, the agreement between experimental (full line) and theoretical (dashed line) values is very good. We should stress that not only the maximal values but also the characteristic decay lengths of both the conductance and the short-range forces match very well. Of course, the agreement strongly relies on the choice of our tip model, in particular on the chemical composition of its outermost atoms.

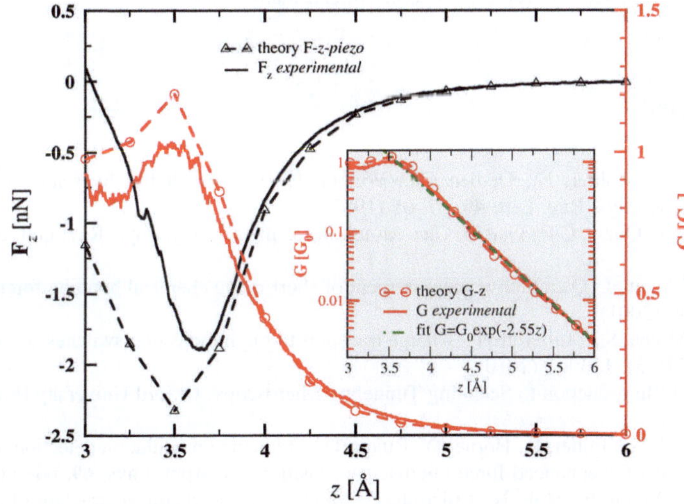


**Fig. 6** Comparison between experimental (*full line*) and theoretical (*dashed line* with symbols) results obtained on atomically sized point contact formed between the tip and the sample. In particular, the figure shows a plot of the short-range force (*black*) and the conductance (*red*) as function of tip–sample distance acquired on a Cu adatom on the Cu(111) surface. More details can be found in Ternes et al. [23]

Finally, we should note that there is a fundamental relationship between the tunneling current and the short-range chemical force, steaming from the fact that both quantities are a function of wavefunction overlap. What is more, it has been shown recently that this relation is driven by the quantum degeneracy of the frontiers orbital of the tip and the surface atoms in contact (for more details, see [32, 36]). This is one of the examples of how simultaneous nc-AFM/STM measurements supported by theory can extend our knowledge about the relations between different quantities at atomic scale.

## 5 Conclusions

We have discussed the theoretical challenges associated with accurately describing simultaneous AFM/STM measurements. Only total energy DFT calculations combined with transport calculations based on a Green's function technique can correctly reproduce the experimental data. We also discussed the mechanism of the interaction between the tip and the sample when a covalent bond is formed. Finally, we showed how the presence of a covalent bond affects the tunneling barrier, surface dipole, and atomic and electronic structure of both the surface and the sample.

**Acknowledgments** This work has been supported by the GAČR no. 14-02079S. I would like to acknowledge very fruitful discussion with R. Pérez, P. Pou, F. Flores, M. Ondráček, P. Hapala, Y. Sugimoto, O. Custance, M. Ternes, F.J. Giessibl and many others.

## References

1. Binning, G., Rohrer, H., Gerber, Ch., Weibel, E.: Surface studies by scanning tunneling microscopy. Phys. Rev. Lett. **49**, 57–61 (1982)
2. Binning, G., Quate, C.F., Gerber, Ch.: Atomic force microscopy. Phys. Rev. Lett. **56–60**, 930 (1986)
3. Lanz, M.A., et al.: Quantitative measurement of short-range chemical bonding forces. Science **291**, 2580 (2001)
4. van der Molen, S.J., Liljeroth, P.: Charge transport through molecular switches. J. Phys. Cond. Matter. **22**(13), 133001 (2010)
5. Chen, C.J.: Introduction to Scanning Tunneling Microscopy. Oxford University Press, Oxford (2008)
6. Albrecht, T.R., Grütter, P., Horne, D., Rugar, D.: Frequency modulation detection using high-Q cantilevers for enhanced force microscope sensitivity. J. Appl. Phys. **69**, 668 (1991)
7. Gross, L., Mohn, F., Moll, N., Liljeroth, P., Meyer, G.: The chemical structure of a molecule resolved by atomic force microscopy. Science **325**, 1110 (2009)
8. Sugimoto, Y., et al.: Chemical identification of individual surface atoms by atomic force microscopy. Nature **446**, 64–67 (2007)
9. Custance, O., Perez, R., Morita, S.: Atomic force microscopy as a tool for atom manipulation. Nat. Nanotechnol. **4**, 803–810 (2009)

10. Giessibl, F.J.: High-speed force sensor for force microscopy and profilometry utilizing a quartz tuning fork. Appl. Phys. Lett. **73**, 3956–3958 (1998)
11. Giessibl, F.J., Pielmeier, F., Eguchi, T., An, T., Hasegawa, Y.: Comparison of force sensors for atomic force microscopy based on quartz tuning forks and length-extensional resonators. Phys. Rev. B **84**, 125409 (2011)
12. Sawada, D., Sugimoto, Y., Morita, K., Abe, M., Morita, S.: Simultaneous measurement of force and tunneling current at room temperature. Appl. Phys. Lett. **94**, 173117 (2009)
13. Loppacher, Ch., Bammerlin, M., Guggisberg, M., Schol, S., Bennewitz, R., Baratoff, A., Meyer, E., Guntherodt, H.-J.: Dynamic force microscopy of copper surfaces-atomic resolution and distance dependence of tip-sample interaction and tunneling current. Phys. Rev. B **62**, 16944 (2000)
14. Tersoff, J., Hamann, D.R.: Theory and application for the scanning tunneling microscope. Phys. Rev. Lett. **50**, 1998–2001 (1983)
15. Bardeen, J.: Tunneling from a many-particle point of view. Phys. Rev. Lett. **6**(2), 57–59 (1961)
16. Chen, C.J.: Tunneling matrix elements in three dimensional space: the derivative rule and the sum rule. Phys. Rev. B **42**, 8841–8857 (1990)
17. Weymouth, A.J., Wutscher, T., Welker, J., Hofmann, T., Giessibl, F.J.: A phantom force induced by the tunneling current, characterized on Si(111). Phys. Rev. Lett. **106**, 226801 (2011)
18. Majzik, Z., Setvín, M., Bettac, A., Feltz, A., Cháb, V., Jelínek, P.: Simultaneous current, force and dissipation measurements on the Si(111) 7 × 7 surface with an optimized qPlus AFM/STM technique. Beilstein J. Nanotechnol. **3**, 249 (2012)
19. Hofer, W.A., Foster, A.S., Shluger, A.L.: Theories of scanning probe microscopes at the atomic scale. Rev. Mod. Phys. **75**, 1287 (2003)
20. Sugimoto, Y., Custance, O., Morita, S., Abe, M., Pou, P., Jelinek, P., Pérez, R.: Topographic and force spectroscopic study of the Sn/Si(111)-($\sqrt{3} \times \sqrt{3}$)R30° surface using atomic force microscopy. Phys. Rev. B **73**, 205329 (2006)
21. Foster, A.S., Barth, C., Shluger, A.L., Reichling, M.: Unambiguous interpretation of atomically resolved force microscopy images of an insulator. Phys. Rev. Lett. **86**, 2373 (2001)
22. Lantz, M.A., Hoffmann, R., Foster, A.S., Baratoff, A., Hug, H.J., Hidber, H.R., Güntherodt, H. J.: Site-specific force-distance characteristics on the NaCl(001) surface: measurements versus atomistic simulations. Phys. Rev. B **74**, 245426 (2006)
23. Ternes, M., Gonzalez, C., Lutz, ChP, Hapala, P., Giessibl, F.J., Jelinek, P., Heinrich, A.J.: Interplay of conductance, force and structural change in metallic point contacts. Phys. Rev. Lett. **106**, 016802 (2011)
24. Chiutu, C., Sweetman, A.M., Lakin, A.J., Stannard, A., Jarvis, S., Kantorovich, L., Dunn, J.L., Moriarty, P.: Precise orientation of a single C60 molecule on the tip of a scanning probe microscope. Phys. Rev. Lett. **108**, 268302 (2012)
25. Guillermet, O., Gauthier, S., Joachim, Ch., de Mendoza, P., Lauterbach, T., Echavarren, A.: STM and AFM high resolution intramolecular imaging of a single decastarphene molecule. Chem. Phys. Lett. **511**, 482–485 (2011)
26. Pérez, R., Payne, M.C., Stich, I., Terakura, K.: Role of covalent tip-surface interactions in noncontact atomic force microscopy on reactive surfaces. Phys. Rev. Lett. **78**, 678–681 (1997)
27. Yurtsever, A., Sugimoto, Y., Tanaka, H., Abe, M., Morita, S., Ondráček, M., Pou, P., Pérez, R., Jelínek, P.: Force mapping on a partially H-covered Si(111)-(7 × 7) surface: influence of tip and surface reactivity. Phys. Rev. B **87**, 155403(1)–155403(10) (2013)
28. Pou, P., Ghasemi, S.A., Jelinek, P., Lenosky, T., Goedecker, S., Perez, R.: Structure and stability of semiconductor tip apexes for atomic force microscopy. Nanotechnology **20**, 264015 (2009)
29. Blanco, J.M., Flores, F., Pérez, R.: STM-theory: image potential, chemistry and surface relaxation. Prog. Surf. Sci. **81**, 403–443 (2006)
30. Caroli, C., Combescot, R., Nozieres, P., Saint-James, D.: Direct calculation of the tunneling current. J. Phys. C **4**, 916–929 (1971)

31. Ondráček, M., Pou, P., Rozsíval, V., González, C., Jelínek, P., Pérez, R.: Forces and currents in carbon nanostructures: are we imaging atoms? Phys. Rev. Lett. **106**, 176101 (2011)
32. Jelínek, P., Ondráček, M., Flores, F.: Relation between the chemical force and the tunnelling current in atomic point contacts: a simple model. J. Phys. Condens. Mat. **24**, 084001 (2012)
33. Blanco, J.M., Gonzalez, C., Jelinek, P., Ortega, J., Flores, F., Perez, R.: First-principles simulations of STM images: from tunneling to the contact regime. Phys. Rev. B **70**, 085405 (2004)
34. Jelinek, P., Svec, M., Pou, P., Perez, R., Chab, V.: Tip-induced reduction of the resonant tunneling current on semiconductor surfaces. Phys. Rev. Lett. **101**, 176101 (2008)
35. Sadewasser, S., Jelinek, P., Fang, C., Custance, O., Yamada, Y., Sugimoto, Y., Abe, M., Morita, S.: New insights on atomic-resolution frequency-modulation Kelvin probe force microscopy imaging on semiconductors. Phys. Rev. Lett. **103**, 266103 (2009)
36. Sugimoto, Y., Ondráček, M., Abe, M., Pou, P., Morita, S., Pérez, R., Flores, F., Jelínek, P.: Quantum degeneracy in atomic point contacts revealed by chemical force and conductance. Phys. Rev. Lett. **111**, 106803 (2013)

# Manipulation of Metal Nanoparticles on Insulating Surfaces

**Clemens Barth**

**Abstract** In this chapter, it will be shown that with nanometer precision metal nanoparticles can be laterally moved on a surface of an insulator by using the AFM tip. After a general description of requirements needed for lateral manipulation experiments, the surface morphology and lateral manipulation experiments are discussed for the particular case *gold nanoparticles on NaCl(001)*. It will be shown that defects of the NaCl(001) surface underneath the nanoparticles have a strong influence on the manipulation characteristics. At the end, promising perspectives in the field of *defect-mediated manipulation of nanoparticles* are described.

**Keywords** Gold metal nanoparticles · Insulating surfaces · NaCl(001) · Vacancies · Lateral manipulation · Noncontact atomic force microscopy (nc-AFM) · Constant height mode imaging · Kelvin probe force microscopy · Heterogeneous catalysis · Nanotribology · Nanofriction

## 1 Introduction and Motivation

Frequency-modulated noncontact atomic force microscopy (FM nc-AFM) is nowadays an indispensable technique in surface science and related research fields. Since its introduction in 1991 by Albrecht et al. [1], any almost atomically flat and clean surface can be imaged with atomic resolution, which helps characterizing the atomic surface structure, surface defects and adsorbates [2–4]. Recent revolutionary work could even show that atomic orbitals [5] and the atomic structure of single molecules can be imaged [6–9]. Apart from the imaging, the strength and everlasting driving force of nc-AFM is the implementation of other methods [4] that has lead to techniques like force spectroscopy (FS) [3], magnetic force microscopy (MFM) [10], electrostatic force microscopy (EFM) [11], and Kelvin probe force microscopy (KPFM) [12, 13].

C. Barth (✉)
CNRS, Aix-Marseille Université, UMR 7325 CINaM, Campus de Luminy, Case 913, 13288 Marseille Cedex 09, France
e-mail: barth@cinam.univ-mrs.fr



P. Moriarty and S. Gauthier (eds.), *Imaging and Manipulation of Adsorbates Using Dynamic Force Microscopy*, Advances in Atom and Single Molecule Machines,
DOI 10.1007/978-3-319-17401-3_6

An important advantage of nc-AFM is that the tip can be used as an active agitator to explicitly manipulate atomic and nanometer-sized objects on surfaces under controlled conditions [3]. As illustrated by a couple of examples in this book (see, e.g., the chapter "Atom Manipulation Using Atomic Force Microscopy at Room Temperature"), atoms and molecules can be moved on the surface, and the vertical and lateral forces exerted on the individual adsorbates by the tip can be measured [14]. Furthermore, single nanoparticles (NPs) [15–17] or even single atoms [18] can be charged at a detection level of a single electron, with interesting applications in nanoelectronics or heterogeneous catalysis.

With respect to the lateral manipulation of nano-objects, metal NPs with a size larger than 10 nm have been quite often considered, in particular to understand fundamental mechanisms involved in the friction of sliding nanocontacts—a hot topic in nanotribology [19, 20]. In the last decade, the friction of sliding NPs was studied as a function of contact area (NP size) [21–24], NP shape, velocity, temperature [25–27], environment (humidity) [25, 26], surface roughness, and chemical functionalization [25, 26]. The most prominent topic, and at the same time the big 'mystery', is friction, which depends on the contact area of nanometer-sized interfaces [19]. The law of friction for macroscopic bodies does not predict a dependence, which is, however, completely contradicted in the nanoworld: For instance, when moving large antimony NPs on a highly oriented pyrolytic graphite (HOPG) surface in air, the frictional force increases with increasing contact area [21]. Other experiments in ultrahigh vacuum (UHV), in which the substrate surface and the antimony NPs were prepared under clean conditions, have shown a 'frictional duality': While the majority of NPs featured a smaller friction in comparison with the friction in air, a quarter of the NPs has shown almost no friction and no contact area dependence [22, 23]. As verified recently by numerical calculations [24], a non-existing friction can be explained by the model of *superlubricity* [28] or *structural lubricity* [29] for nanocontacts build of two surfaces with incommensurate lattices.[1]

However, the precise mechanisms involved in the contact area-dependent friction are still under debate: For instance, after the structure (amorphous vs. crystalline NPs), shape, and relative orientation of the NPs could be 'excluded' as possible explanations for Sb/HOPG [23], it was anticipated that impurities like molecules [22, 23] or even small parts of the NP [24] are intercalated in the NP–surface interface, which may influence the properties of friction. Apart from the latter explanations, an other one has been discussed in literature [22]: The friction of NPs is influenced by defects like interfacial defects or grain boundaries located at the NP facets.

Although defects must not necessarily be the main origin for the contact area-dependent friction of large NPs, their important role is, however, not far-fetched. In particular, defects are of utmost importance when considering clean NP/surface systems with a well-defined crystalline interface and NPs of very small size (a few nanometers) in, e.g., heterogeneous (model) catalysis [30–33]. In this research

[1]Either one or both surfaces are amorphous or both have different lattice dimensions. In the case of SbNPs on HOPG, the lattice mismatch is around 4 %.

field, it is known for a long time that surface defects (e.g., vacancies, impurities, steps) are the nucleation sites for metal NPs. If platinum, palladium, or gold atoms are vacuum evaporated onto oxide surfaces, the atoms are mobile and form NPs at such defects. If the AFM tip tries now to move a NP, a couple of possible phenomena may arise: For instance, the defect remains at its surface site and energy is needed to dissociate the defect–NP complex, which costs energy. In a second scenario, the defect (e.g., vacancy) moves with the moving NP and also here an energy barrier must be overpassed, e.g., to start the movement of the NP. An influence of defects certainly has a high impact on the manipulation characteristics but also on the friction of NPs.

Although the role of defects at the NP–surface interface deserves advertence due to its importance as recently shown [34], it has been almost neglected in the manipulation of NPs. In fact, it forms a new domain in the manipulation and friction of NPs, which can be called *defect-mediated manipulation*. The defect-mediated manipulation of NPs opens many promising future perspectives in heterogeneous model catalysis as it will be discussed at the end of this chapter.

In this chapter, we focus on a case study, which was recently presented in Ref. [34]. It is a pioneer work, which documents a controlled lateral manipulation of small gold NPs on clean and crystalline NaCl(001) surfaces. We discuss first general criteria needed for accomplishing precise manipulation experiments and focus on the clean alkali halide surfaces, related surface defects, and supported metal NPs. The manipulation itself is discussed in the following section, whereas the last section summarizes and comments future perspectives.

## 2 Criteria for Manipulation Experiments

To understand all mechanisms involved in the lateral manipulation of NPs on surfaces, experiments should meet the following criteria:

- Experiments have to be done under extreme clean conditions (UHV) in order to rule out possible contaminants from the environment, in particular at the NP–surface interface [22]. Experiments should be therefore accomplished under UHV conditions.
- The surface and NPs should have a well-defined structure (crystalline) with known properties at the interface (epitaxy, etc.) and of the NP (shape, facets, etc.).
- The surfaces and NPs should remain clean within the period of experimental observations—the surface and NPs should be rather inert in UHV.
- The number of different types of surface defects should be small, and the properties of the few defects should be known (type of defect, surface site, electronic properties, etc.). Furthermore, the defects should be stable within the period of experimental observations.
- The surface should give an easy access for atomic resolution imaging in the nc-AFM mode (for determining the precise lattice orientation of NPs and surface) and KPFM (for analyzing the charge state, charge transfer, etc. of NPs) such that a manipulation of NPs can be accomplished.

In principle, apart from surfaces like HOPG, the surfaces of ionic insulators are ideal candidates. For instance, the (111) surfaces of all three earth alkali halides ($CaF_2$, $SrF_2$, and $BaF_2$) and also the (001) surfaces of alkali halides (NaCl, KBr, KCl, LiF, etc.) almost fully meet the criteria from above. We discuss this for alkali halides in the following.

Alkali halides have been used for extensive studies in condensed matter physics [35], and the crystal structure, defects, doping, ionic conductivity, surface reactivity, etc. have been addressed several decades ago [36, 37]. The (001) surface morphology has been studied in dependence on surface diffusion and evaporation [38–41] and also on impurities [42–44] by the gold decoration method starting from the end of the 1950s [45]. With respect to FM nc-AFM [1], which is the high-resolution imaging technique of choice in this case, alkali halide surfaces have become *standard model surfaces* for many applications such as atomic structure analysis [46–54], charge identification [55–57], indentation [58] as well as adsorbed molecules [59, 60], self-assembled molecules [61–70] but also metal NPs [17, 69, 71–76].

An alkali halide (001) surface can be easily prepared by cleaving a single crystal along the (001) cleavage plane in UHV and by following in situ annealing [55, 77, 78]. The same surface can also be obtained by evaporating a small amount of the alkali halide material onto a clean metal surface like silver (001) (see Ref. [79] and references therein). After the preparation, the surface exhibits large atomically and perfectly stoichiometric flat terraces—ideal conditions for nc-AFM and KPFM, especially with respect to atomic resolution imaging and the imaging of supported NPs. Most importantly, the surfaces are quite inert in UHV such that measurements can be accomplished within a week during which the surface remains clean. Furthermore, the number of possible defects, which are stable during the acquisition time in AFM experiments, is quite limited: Apart from impurities, only three types of vacancies exist: cation and anion vacancies, and $F^0$ centers, which are anion vacancies with a localized electron inside.

With respect to the NP material, any kind of metal can be used since all metals form three-dimensional NPs on alkali halide surfaces [71, 73, 75]. Furthermore, NPs can be prepared in a very clean manner under UHV conditions, e.g., by evaporating the metal onto the surface. Last but not least, if the substrate temperature is sufficiently high during the growth of NPs, most of them are crystalline and exhibit a well-defined interface with the surface [30]. A good NP material for manipulation experiments is gold because it is quite inert and well studied. Furthermore, gold NPs are relevant in heterogeneous catalysis [80].

## 3 The (001) Surfaces of Alkali Halides and Supported Metal NPs

Once an alkali halide crystal surface is prepared by UHV cleavage and the crystal is put into its equilibrium charge state by annealing (e.g., at $T \approx 100$ °C for some hours), typical features of the (001) surface can be imaged by nc-AFM. Topography

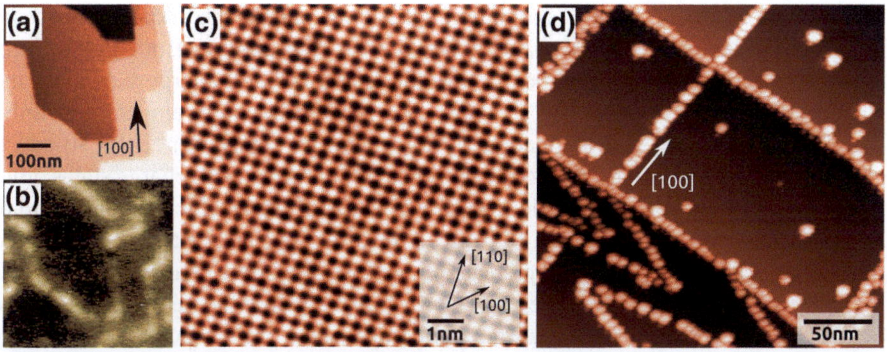


**Fig. 1** The (001) surfaces of alkali halides and supported metal AuNPs. **a** KBr(001) surface with large atomically flat terraces and mono-atomic high steps (topography image, crystal preparation: air cleavage and UHV annealing at $T$ = 200 °C for some hours). The corresponding Kelvin image **b** reflects the net negative surface charge of the Debye-Frenkel layer. Round steps with a high density of cation vacancies exhibit a stronger bright contrast than the almost perfect ⟨100⟩ steps, which only have a low density of cation vacancies. **c** Topography image acquired on NaCl(001) at low temperature (5 K) with an Omicron low-temperature AFM/STM (QPlus sensor based), directly after the cleavage of the crystal in UHV. In general, no defects can be found on the flat terraces. **d** AuNPs on the NaCl(001) surface (crystal preparation: UHV cleavage and following annealing at $T$ = 80 °C for 2 h, AuNP preparation: 0.32 monolayer nominal thickness, $T$ = 190 °C growth temperature.). Images **a** and **b** are a courtesy of A. Hinaut, D. Martrou, S. Gauthier (CEMES-CNRS, Toulouse, France). Images **c** and **d** are unpublished images of the author (Color figure online)

images at large scale (Fig. 1a) show atomically flat and up to 150-nm-wide terraces, which are divided by mono-atomic high steps. In most cases, images with atomic resolution show no *detectable* defects on the terraces (e.g., ad-ions, vacancies, and impurities), they rather show the stoichiometric ionic lattice of the surface (Fig. 1c), which is also supported by metal deposition experiments that show an extremely low NP density in such regions (Fig. 1d).

Note that the lack of 'visible' defects is always related to the timescale of the image acquisition. Observing no defects by nc-AFM does not necessarily mean that there are no *mobile* surface defects. For instance, at room temperature, the diffusion of vacancies on terraces is much faster in comparison with the AFM scanning speed [72, 81]—vacancies are therefore 'invisible' for the AFM. Furthermore, a $F^0$ center (chlorine vacancy + electron) does not significantly change the image contrast because the electron in the vacancy does not disturb the alternating electrostatic potential of the surface ions [82]. In other words, such defects are also not 'visible' for the AFM, and they can be only indirectly 'detected' by, e.g., a decoration with metal [83].

With respect to the NaCl(001) bulk surface, a fundamental surface property of ionic materials in condensed matter physics describes defects on the surface: Due to the different free formation energies needed for the creation of an anion and cation vacancy at the surface, a double layer is created (*Surface double layer* or *Debye-Frenkel layer*) [84, 85]. One type of vacancy is located on the surface, whereas the other one is stabilized in a layer below the surface, in a so-called *space charge layer*

[86–90]. Note that a cation and anion vacancy changes the local electrostatic potential producing a negative and positive site, respectively. Both layers form therefore a dipole, which modifies in turn the free formation energy balancing the electrostatics of the crystal. This *intrinsic case* is valid for high temperatures only, whereas at relatively low temperatures, like at room temperature, the so-called *extrinsic case* predominates due to the presence of impurities with charge states other than +1 or −1 [87]. Such impurities can be $Ca^{2+}$ or $Mg^{2+}$ ions which can be found in any crystal of even highest purity. Because of the different charge state with respect to the $Na^{+}$ ions, cation vacancies are created in order to conserve charge neutrality. At thermal equilibrium, a Debye-Frenkel layer is built such that the net negative charge of the cation vacancies on the surface is compensated by the positive impurities below the surface. The cation vacancies find their energetic equilibrium position at low-coordinated surface sites of the steps [72, 91, 92]. Important to note is that a very small impurity amount of only a few ppm is already sufficiently large to influence the Debye-Frenkel layer and in turn the net surface charge [87]. Furthermore, it is very difficult to include negative impurities with a charge state other than −1 into the host system, which would lead to positive anion vacancies on the surface and therefore to a positive net surface charge. Therefore, because only positive impurities are present the net surface charge is always negative.

A work supplying direct evidence of the Debye-Frenkel layer was published some years ago [55], and following work could approve the experimental observations [56, 57]. The net negative surface charge of the cation vacancies can be directly detected by KPFM because such negative defects modify the local electrostatic potential of the surface and in turn the Kelvin voltage (Fig. 1b): With respect to the neutral and stoichiometric NaCl terraces (dark), bright patches at the steps corresponding to negative surface sites can be observed in Kelvin images [78]. At such positions, nc-AFM imaging with atomic resolution reveals single atomic-sized defects at kinks and corners of steps, which are the negative cation vacancies [55]. Note that the vacancies are electrostatic point sources, which leads to an imaging of the tip apex. This explains why the bright patches in the Kelvin image (Fig. 1b) have a size of some nanometers.

## 4 Manipulation of AuNPs on NaCl(001)

A typical manipulation experiment of gold NPs with about 2000 atoms (mean diameter ~5 nm) is shown in Fig. 2. The detuning $\Delta f$ image (a) shows single AuNPs in a dark contrast (more negative $\Delta f$ values), which decorate mono-atomic high steps oriented toward the $\langle 001 \rangle$ and $\langle 011 \rangle$ surface directions. During a manipulation experiment, the NP in a green color (Fig. 2a) could be detached from a step and moved along a path of 12 ± 1 nm onto a terrace site (Fig. 2b). In following experiment, the NP could be moved again by 15 ± 1 nm from the first terrace site (Fig. 2d) and again to another site by 16 ± 1 nm (Fig. 2e).

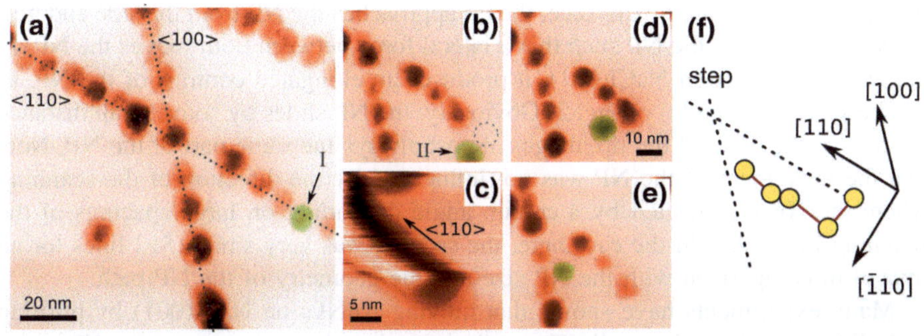


**Fig. 2** Experimental constant height images presenting several manipulation steps of AuNPs on NaCl(001) (Au coverage: 0.64 monolayers). **a** Decoration of NaCl steps by AuNPs (NP–NP distance: 8.4 nm, NP density: $4.6 \times 10^{11}$ NPs/nm$^2$). **b** Image after a single NP (*green*) was detached from the step. The *dashed circle* represents the position of the manipulated NP in image **a**. **c** Controlled manipulation of the NP by sliding. **d** The result after the movement. **e** Controlled manipulation by pushing in a spectroscopy step. **f** Schematic of the manipulation steps. Adapted from Ref. [34] (Color figure online)

The manipulation steps were achieved by either pushing the NP during a spectroscopy step (Fig. 2d–e) or by sliding the NP during imaging in the nc-AFM mode (Fig. 2b–d). In the first method, the scanning is switched off and the tip is placed on or very close to the NP (Fig. 3a). The tip–sample distance is then gradually decreased such that at a close NP-tip distance, the NP moves on the surface due to a sufficiently large lateral force between the foremost atoms of the tip and the NP. In the second method, the tip is scanned in the regular raster scan pathway fashion during imaging (Fig. 3b) [93]. In particular, the constant height mode is the scanning mode of choice since it allows a fast imaging of the surface, minimizes the tip–surface convolution effect, and considerably reduces the drift of the piezo scanner [94–96]. During a manipulation, the tip–surface distance is reduced from image to image by choosing more negative values for the mean

**Fig. 3** Protocols used for the manipulation of AuNPs: **a** A nanoparticle is moved by pushing during a spectroscopy step. **b** The NP is moved by sliding during scanning (Color figure online)

detuning $\Delta f$. At a specific distance, the tip approaches the NP from the side and gets sufficiently close to the NP such that a lateral force between both moves the NP. In the case of AuNPs on NaCl(001), it produces an elongated contrast of the NP, as can be seen in image (c) of Fig. 2. Obviously, the NP slides by some small distance in front of the tip, and the tip images almost always the same part of the NP. Note that the movement of the NP was not influenced by the direction of the scanning where the NP is deflected by a certain angle depending on large spacings of the scanning lines [93]. In the case here, the spacing was very small (512 lines for an image) in comparison with the NP size and the geometry of the NP path.

Many experiments have shown that moving AuNPs on NaCl(001) by pushing and sliding works quite well. They all show that the movement of a NP, once detached from a step, is in fact anisotropic—the NP preferentially moves along the $\langle 011 \rangle$ surface direction on the flat terrace. Surprisingly, the NP is indeed stable on the terrace and never moves from a terrace position at room temperature. This is an important observation because it is known that NPs on the perfect NaCl(001) surface are mobile [34, 83, 97, 98]. From this, it follows that the stable NP position on terraces must be related to an anchorage of the NP at surface defects and that a defect probably moves at the same time when a NP moves.

Density functional theory (DFT) calculations accomplished with pyramidal-shaped NPs with 20 and 100 atoms could show the important role of defects (Fig. 4). Since the number of defect types is limited in the case of alkali halides (see Sect. 3), a NP was modeled such that it sits either on a $F^0$ center, chlorine ($V_{Cl^-}$), or sodium ($V_{Na^+}$) vacancy of the NaCl(001) surface. Indeed, for all configurations, the adsorption energies are quite high ($V_{Cl^-}$: 2.1 eV and $V_{Na^+}$: 2.5 eV), which supports

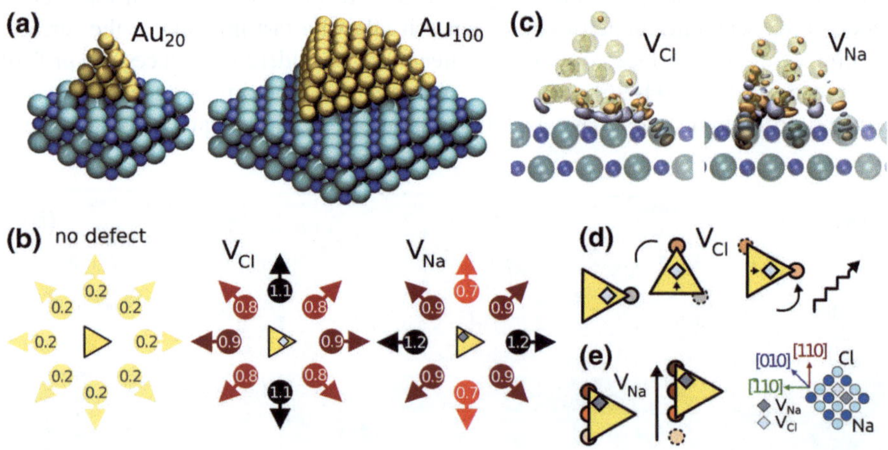


**Fig. 4** **a** The $Au_{20}$NPs (*left*) and $Au_{100}$NPs (*right*) used for DFT calculations. **b** Calculated diffusion barriers for some $Au_{20}$NP-vacancy structures in eV. The $\langle 110 \rangle$ directions are set to run horizontally and vertically. *left* $Au_{20}$NP without defect, *middle* $Au_{20}$NP on a Cl vacancy and *right* $Au_{20}$NP on a Na vacancy. **c** Differential charge density images for $Au_{20}$NP on a Cl vacancy (*left*) and on a Na vacancy (*right*). **d** and **e** Schematic representation of the preferred diffusion modes for the $Au_{20}$NP on a Cl vacancy (**d**) and a Na vacancy (**e**). Adapted from Ref. [34] (Color figure online)

the assumption that a defect is needed for stabilizing a NP on the terrace. Note that independently, if a real $V_{Cl^-}$ vacancy or a $F^0$ center is modeled, the anion vacancy is always empty (no gold atom inside) and both vacancies have the same electronic configuration due to the large reservoir of electrons inside the NP.

Not only the adsorption energies but also the diffusion energies are higher (Fig. 4b). The surprising result is that it makes a clear difference if the NP sits on a $V_{Cl^-}$ or $V_{Na^+}$ vacancy, independently on the size of the NP: In the first case, one of the edges of the NP bounds to a surface $Cl^-$ ion with the $V_{Cl^-}$ vacancy next to it (Fig. 4c, left). The whole NP is tilted by 10° toward the $Cl^- - V_{Cl^-}$ vacancy complex so that the NP then moves by pivoting around this bonded corner. Always when the NP rotates, the $V_{Cl^-}$ vacancy jumps to a neighboring site at the same time, allowing the NP to bond to another $Cl^-$ ion (Fig. 4d). Although in this case, the direction is along one of the {001} family of surface directions, the NP can principally also move along the other ones—none of the {001} directions is preferred and the NP can move in any direction by taking a zigzag path.

On the contrary, when the NP sits on a $V_{Na^+}$ vacancy, the movement of the NP is highly anisotropic. The defect is located in the middle of one of the NP edges, which is aligned on top of a $Cl^-$ row along the ⟨011⟩ surface direction (Fig. 4b, right). The NP is again slightly tilted by 5.1° toward the edge such that the NP bounds via the edge to the surface. The NP moves with the $V_{Na^+}$ vacancy, and the movement is directed along the ⟨011⟩ direction of the row of $Cl^-$ ions (Fig. 4e). On the contrary, the movement along the other ⟨011⟩ directions is less preferred due to the break of symmetry by the triangular shape of the NP and its (111) epitaxy with respect to the (001) orientation of the NaCl(001) surface (see diffusion energies in Fig. 4b, right).

It can be concluded that a NP needs a defect for getting stabilized on the terrace. More importantly, with respect to the manipulation experiments shown in Fig. 2, the defect probably was a cation vacancy. This is in perfect agreement with the general picture of defects on NaCl(001) that the steps have indeed many cation vacancies at the kinks, as discussed in Sect. 3.

A very important question is whether the ⟨011⟩ direction is always the preferred one. In other words, are there NPs which are attached at anion vacancies? The manipulation of NPs already anchored on the flat terraces directly after the growth of the NPs is sometimes different: Such NPs could be moved almost perfectly along the ⟨001⟩ direction as shown in Fig. 5a. In other experiments, the direction of the NP movement is not aligned along a simple surface direction like ⟨001⟩ or ⟨011⟩, as it can be seen in the experiment shown in Fig. 5b. The NP moved rather in a zigzag path.

Without any direct proof, it can be anticipated that such NPs are possibly anchored at $F^0$ centers, which is yet again in agreement with the general picture of the clean surface: Because the NaCl crystals are annealed at relatively high temperatures after their cleavage, most of the surface cation vacancies move to the steps due to their high mobility and get trapped at the steps [72]. However, for $F^0$ centers, the diffusion energy is much higher (0.9 eV [34]), and it could be that a few of them are present and stable on the surface even after the surface preparation by cleavage and annealing of the crystals. If so, they serve as nucleation sites for metal atoms, which then form NPs at $F^0$ centers as recently described in detail [83].

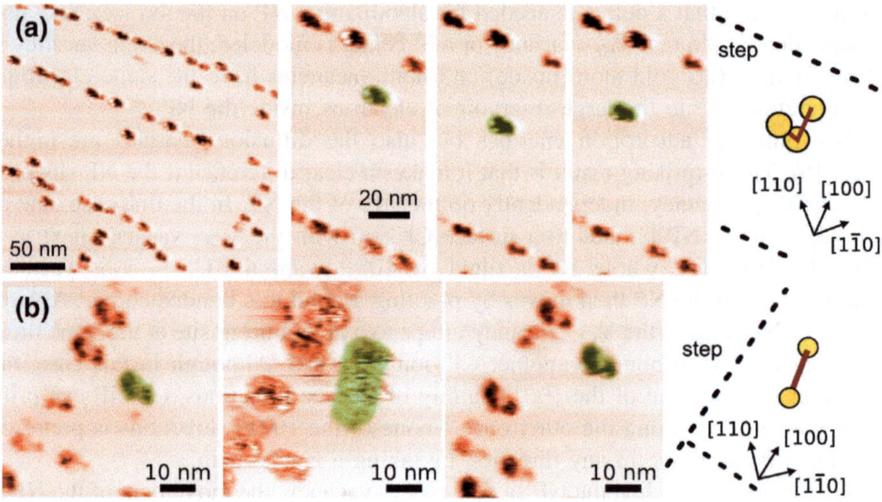


**Fig. 5** Manipulation of two AuNPs (green) on NaCl(001) (Au coverage: 0.3 monolayers). **a** The NP could be moved almost close along the $\langle 100 \rangle$ surface direction. **b** The NP moved along a complex surface direction exhibiting a zigzag path. The images of both experiments, (**a**) and (**b**), always show the situation before and after the manipulation. Experiment (**a**) was adapted from Ref. [34]. Experiment (**b**) are unpublished results of the author. Experiment (**b**): 60 × 60 nm$^2$, $\Delta f = -20$, $-42$, $-21$ Hz, $f_0$ = 275 kHz, $k$ = 28 N/m, $v_{Scan} \approx$ 20 Hz, $A_{pp}$ = 15 nm (Color figure online)

# 5 Conclusion

The manipulation experiments with AuNPs on NaCl(001) demonstrate that metal NPs can be successfully moved on an insulating surface with already a standard UHV AFM, which works at room temperature.[2] This work *invites* studying the manipulation characteristics of any type of metal NP on well-defined, almost atomically flat and clean surfaces.

The manipulation experiments show that AuNPs need a defect on the NaCl terraces for their stabilization and that in most cases, the direction of manipulation is anisotropic, which is related to the type of defect below the NP. In fact, the preferred direction of NPs can be used to identify the defect underneath. Although the picture from Sect. 4 is coherent and in perfect agreement with, e.g., the Debye-Frenkel layer that describes the type of defect at NaCl steps, there is no direct evidence for the existence of the cation and anion vacancies below the NPs—the NPs were not removed from the surface to see what is left behind on the surface. An unambiguous identification of the defects after or before a manipulation experiment is, however, of importance to prove the model.

[2]The UHV AFM used for the experiments in Ref. [34] is a standard commercial Omicron UHV AFM/STM [99], controlled with a SCALA computer system. The detection of the detuning $\Delta f$ has been improved using a digital demodulator from NanoSurf AG [100].

Experiments that prove the existence of defects are not only needed for the particular case AuNPs/NaCl(001) but are relevant for any kind of defect-mediated manipulation experiment. A general protocol for a complete characterization of the manipulation characteristics of one single NP can be formulated containing the following steps:

1. Characterizing the manipulation characteristics of the NP
2. Gently lifting the NP from the surface
3. Imaging and identifying the defect that is left behind

In the first manipulation step, a NP from, e.g., a NaCl step is moved onto a terrace site. When detached from the step, the NP should be repeatedly moved on the terrace to observe a trend in the direction of movement of the NP, e.g., a possible anisotropic movement. This manipulation step can be easily accomplished at room and low temperature. For the second manipulation step, a low-temperature AFM is needed: Not only a high precision is given for accomplishing this manipulation step, but it is also guaranteed that the exposed defect does not vanish from its site (3rd step), which might happen at relatively high temperatures as in the case of NaCl(001).

The most difficult part of the protocol is the lift-off of the NP, which needs to be picked up by the AFM tip (2nd step). Although some experiments in air demonstrate that NPs can be picked up in principle, there is, so far, no example in literature demonstrating this also in vacuum. Merely, atoms and molecules could be picked up from the surface [101], which let hope that lifting a NP should be possible, too. The manipulation step asks for an attentive execution of the experiment and of course also some experimental luck.

The last manipulation step is easy to accomplish since it only concerns the basic imaging of the defect with atomic resolution. The atomic resolution imaging can be done with the NP at the tip apex, otherwise one desorbs the NP from the tip by, e.g., voltage pulses. The atomic resolution imaging should be done systematically such that the defect can be unambiguously identified afterward, e.g., with help of assisting theory [82]. Either distance-dependent imaging [54, 94, 102, 103], 2D [14, 104] or even true 3D force spectroscopy [105, 106] should be done to fully characterize the defect contrast.

## 6 Future Perspectives

With respect to the manipulation experiments from above (Sect. 4) and manipulation experiments of single atoms and molecules that can be nowadays done with high precision (e.g., see Refs. [14, 18]), several appealing experiments could be done in future. For instance, an interesting experiment would be the manipulation of NPs as a function of surface structure and in particular as a function of defect type and size. This results from the simple point of view that a large number of interesting surfaces can be found in *surface science*, which all differ in their respective atomic surface structure but also in their surface defects.

Promising surfaces are the (001) surfaces of alkali halides, where vacancies with different sizes can be found and can be explicitly created (pits). For instance, because the dimension of the unit cell varies from material to material (LiF: $a_{LiF} = 4.03$ Å, NaCl: $a_{NaCl} = 5.64$ Å, KBr: $a_{KBr} = 6.60$ Å, RbBr: $a_{RbBr} = 6.85$ Å), the size of the cation vacancy also varies. At a specific lattice size, the vacancy is sufficiently large such that a metal atom of the NP can fill the vacancy [72]. For instance, in the case of gold on KBr(001) theory predicts that a gold atom fills a $F^0$ vacancy [72], which is not the case on NaCl(001) [34] and probably also on LiF (001). An other way to size the dimension of the defect underneath a NP is to make use of electron desorption, which creates nanometer-sized pits on the surface [74, 107]. Pits with a size ranging from the atomic size (single vacancy) up to small pits with a surface area still smaller than the contact area with the NP could be created. Last but not least, a few separated vacancies underneath a NP could be considered. A key substrate surface is the nanostructured (001) Suzuki surface of, e.g., NaCl doped with divalent metal impurity cations (e.g., $Cd^{2+}$ or $Mg^{2+}$) [51, 108]. The Suzuki surface exhibits a high density of vacancies and impurities with a mean next-neighbor distance of $\sim 8$ Å [54] so that NPs with a size of only a few nanometers already cover two cation vacancies and more [69].

The movement of a NP, which covers a filled vacancy or pit, is certainly different in comparison with the movement of a NP anchored at an unfilled vacancy or pit: If a gold atom is inside a vacancy or pit, it needs to be moved with the vacancy and the NP, or the defect–NP complex gets dissociated. Such scenarios cost probably much more energy with respect to the unfilled case. In the case of several single vacancies (Suzuki surface), a concerted movement of the NP and the defects is needed to move the ensemble together, which probably also costs more energy. In future, manipulation experiments on alkali halide surfaces could be done to clarify the effect of the vacancy size.

Alkali halide surfaces form one class of substrate surfaces. In fact, they are suitable *model surfaces* for the manipulation of NPs. Other surfaces like oxide surfaces are more relevant in, e.g., heterogeneous catalysis. Today, a precise knowledge of almost all relevant defects exists, in particular with respect to oxygen vacancies. This concerns the MgO(001) [109], $TiO_2$(110) [110–112], $CeO_2$(111) [113], and $Al_2O_3$ surface [114], for instance. In most cases, thin oxide films with fascinating morphologies, atomic structures, reconstructions, and defect structures can be prepared in a controlled manner [115–119], with an easy access for nc-AFM and KPFM imaging [4, 120]. If one succeeds with the manipulation methods from Sect. 5, a dream in heterogeneous model catalysis becomes true: the morphology of single NPs, charge transfer processes between the oxide support and single NPs but also reactivity related phenomena like adsorption and contamination at single NPs (see Ref. [121]) can be studied as a function of surface site.

**Acknowledgments** The author is particularly grateful for a part of the experimental work done by G. Cabailh. He expresses his very great appreciation to T. Hyninnen and A.S. Foster for the numerical simulations and fruitful collaboration they have had together in the recent past. Stimulating discussions with C.R. Henry and S. Gauthier are also greatly appreciated. The author acknowledges the European Science Foundation for financial support through the FANAS project NOMCIS and the French agency for Research (Agence Nationale pour la Recherche, ANR) for financial support through the programs CANA and MISS.

# References

1. Albrecht, T.R., Grütter, P., Horne, D., Rugar, D.: Frequency modulation detection using high-Q cantilevers for enhanced force microscope sensitivity. J. Appl. Phys. **69**, 668 (1991)
2. Morita, S., Wiesendanger, R., Meyer, E.: Noncontact Atomic Force Microscopy, vol. 1. Springer, Berlin (2002)
3. Morita, S., Giessibl, F., Wiesendanger, R.: Noncontact Atomic Force Microscopy, vol. 2. Springer, Berlin (2009)
4. Barth, C., Foster, A.S., Henry, C.R., Shluger, A.L.: Recent trends in surface characterization and chemistry with high-resolution scanning force methods. Adv. Mater. **23**, 477–501 (2011)
5. Giessibl, F.J., Hembacher, S., Bielefeldt, H., Mannhart, J.: Subatomic features on the silicon (111)-(7 × 7) surface observed by atomic force microscopy. Science **289**, 422 (2000)
6. Gross, L., Mohn, F., Moll, N., Liljeroth, P., Meyer, G.: The chemical structure of a molecule resolved by atomic force microscopy. Science **325**, 1110–1114 (2009)
7. Gross, L., Mohn, F., Moll, N., Meyer, G., Ebel, R., Abdel-Mageed, W.M., Jaspars, M.: Organic structure determination using atomic-resolution scanning probe microscopy. Nat. Chem. **2**(10), 821–825 (2010)
8. Guillermet, O., Gauthier, S., Joachim, C., de Mendoza, P., Lauterbach, T., Echavarren, A.: STM and AFM high resolution intramolecular imaging of a single decastarphene molecule. Chem. Phys. Lett. **511**, 482–485 (2011)
9. Gross, L., Mohn, F., Moll, N., Schuler, B., Criado, A., Guitián, E., Peña, D., Gourdon, A., Meyer, G.: Bond-order discrimination by atomic force microscopy. Science **337**(6100), 1326–1329 (2012)
10. Wiesendanger, R., Schwarz, A.: Magnetic sensitive force microscopy. Nano Today **3**, 28–39 (2008)
11. Girard, P.: Electrostatic force microscopy: principles and some applications to semiconductors. Nanotechnology **12**, 485–490 (2001)
12. Melitz, W., Shen, J., Kummel, A.C., Lee, S.: Kelvin probe force microscopy and its application. Surf. Sci. Rep. **66**, 1–27 (2011)
13. Sadewasser, S. Glatzel, T.: Kelvin Probe Force Microscopy: Measuring and Compensating Electrostatic Forces. In: Ertl, G., Luth, H., Mills, D. (eds.) Surface Science, Ist edn. Springer, Berlin (2012)
14. Ternes, M., Lutz, C.P., Hirjibehedin, C.F., Giessibl, F.J., Heinrich, A.J.: The force needed to move an atom on a surface. Science **319**, 1066 (2008)
15. Stomp, R., Miyahara, Y., Schaer, S., Sun, Q., Guo, H., Grutter, P., Studenikin, S., Poole, P., Sachrajda, A.: Detection of single-electron charging in an individual InAs quantum dot by noncontact atomic-force microscopy. Phys. Rev. Lett. **94**, 056802 (2005)
16. Cockins, L., Miyahara, Y., Bennett, S.D., Clerk, A.A., Studenikin, S., Poole, P., Sachrajda, A., Grutter, P.: Energy levels of few-electron quantum dots imaged and characterized by atomic force microscopy. Proc. Natl. Acad. Sci. **107**(21), 9496 (2010)
17. Tekiel, A., Miyahara, Y., Topple, J.M., Grütter, P.: Room-temperature single-electron charging detected by electrostatic force microscopy. ACS Nano **7**, 4683–4690 (2013)

18. Gross, L., Mohn, F., Liljeroth, P., Repp, J., Giessibl, F.J., Meyer, G.: Measuring the charge state of an adatom with noncontact atomic force microscopy. Science **324**, 1428 (2009)
19. Schirmeisen, A., Schwarz, U.D.: Measuring the friction of nanoparticles: a new route towards a better understanding of nanoscale friction. Chem. Phys. Chem. **10**(14), 2373–2382 (2009)
20. Dong, Y., Li, Q., Martini, A.: Molecular dynamics simulation of atomic friction: a review and guide. J. Vac. Sci. Technol. A **31**(3), 030801 (2013)
21. Ritter, C., Heyde, M., Stegemann, B., Rademann, K., Schwarz, U.D.: Contact-area dependence of frictional forces: moving adsorbed antimony nanoparticles. Phys. Rev. B **71** (8), 085405 (2005)
22. Dietzel, D., Ritter, C., Mönninghoff, T., Fuchs, H., Schirmeisen, A., Schwarz, U.D.: Frictional duality observed during nanoparticle sliding. Phys. Rev. Lett. **101**, 125505 (2008)
23. Dietzel, D., Mönninghoff, T., Herding, C., Feldmann, M., Fuchs, H., Stegemann, B., Ritter, C., Schwarz, U.D., Schirmeisen, A.: Frictional duality of metallic nanoparticles: influence of particle morphology, orientation, and air exposure. Phys. Rev. B **82**(3), 035401 (2010)
24. Brndiar, J., Turanskỳ, R., Dietzel, D., Schirmeisen, A., Štich, I.: Understanding frictional duality and bi-duality: Sb-nanoparticles on HOPG. Nanotechnology **22**(8), 085704 (2011)
25. Mougin, K., Gnecco, E., Rao, A., Cuberes, M.T., Jayaraman, S., McFarland, E.W., Haidara, H., Meyers, E.: Manipulation of gold nanoparticles: influence of surface chemistry, temperature, and environment (Vacuum versus ambient atmosphere). Langmuir **24**, 1577–1581 (2008)
26. Darwich, S., Mougin, K., Rao, A., Gnecco, E., Jayaraman, S., Haidara, H.: Manipulation of gold colloidal nanoparticles with atomic force microscopy in dynamic mode: influence of particle–substrate chemistry and morphology, and of operating conditions. Beilstein J. Nanotechnol. **2**(1), 85–98 (2011)
27. Tripathi, M., Paolicelli, G., D'Addato, S., Valeri, S.: Controlled AFM detachments and movement of nanoparticles: gold clusters on HOPG at different temperatures. Nanotechnology **23**(24), 245706 (2012)
28. Hirano, M., Shinjo, K.: Atomistic locking and friction. Phys. Rev. B **41**, 11837–11851 (1990)
29. Müser, M.H.: Structural lubricity: role of dimension and symmetry. EPL (Europhys. Lett.) **66** (1), 97–103 (2004)
30. Henry, C.R.: Surface studies of supported model catalysts. Surf. Sci. Rep. **31**, 231–325 (1998)
31. Libuda, J., Freund, H.-J.: Molecular beam experiments on model catalysts. Surf. Sci. Rep. **57**, 157–298 (2005)
32. Heiz, U., Landman, U.: Nanocatalysis. Springer, Berlin (2007)
33. Freund, H.-J.: Model studies in heterogeneous catalysis. Chem. A Eur. J. **16**(31), 9384–9397 (2010)
34. Hynninen, T., Cabailh, G., Foster, A.S., Barth, C.: Defect mediated manipulation of nanoclusters on an insulator. Sci. Rep. **3**, 1270 (2013)
35. Hayes, W., Stoneham, A.M.: Defects and Defect Processes in Nonmetallic Solids. Wiley-Interscience, New York (1985)
36. Seitz, F.: Color centers in alkali halide crystals. Rev. Mod. Phys. **18**, 384–408 (1946)
37. Seitz, F.: Color centers in alkali halide crystals. II. Rev. Mod. Phys. **26**, 7–94 (1954)
38. Bethge, H.: Electron microscopic studies of surface structures and some relations to surface phenomena. Surf. Sci. **3**, 33–41 (1964)
39. Bethge, H., Keller, K.W.: Evaporation rate of NaCl in a vacuum. J. Cryst. Growth **23**, 105–112 (1974)
40. Höche, H., Bethge, H.: Morphological changes of the Monoatomic (100) NaCl cleavage structure due to edge self-diffusion: 1. Exp. J. Cryst. Growth **33**, 246–254 (1976)
41. Bethge, H., Heydenreich, J.: Elektronenmikroskopie in der Festkörperphysik. Springer, Berlin (1982)

42. Bassett, G.A., Yacamán, M.J.: Surface aggregation of impurity in manganese-doped alkali halide crystals. Thin Solid Films **31**, 375–384 (1976)
43. Yacamán, M.J., Hirth, J.P.: Effects of surface imperfections on the nucleation of gold on sodium chloride. Thin Solid Films **38**, 215–223 (1976)
44. Grange, G.: Formation of the suzuki phase on the solid-liquid interface of czochralski grown crystals. Surf. Sci. **105**, 265–274 (1981)
45. Bassett, G.A.: A new technique for decoration of cleavage and slip steps on ionic crystal surfaces. Philos. Mag. **3**, 1042–1045 (1958)
46. Bammerlin, M., Lüthi, R., Meyer, E., Baratoff, A., Lü, J., Guggisberg, M., Loppacher, C., Gerber, C., Güntherodt, H.-J.: Dynamic SFM with true atomic resolution on alkali halide surfaces. Appl. Phys. A **66**, S293–S294 (1998)
47. Bennewitz, R., Schär, S., Gnecco, E., Pfeiffer, O., Bammerlin, M., Meyer, E.: Atomic structure of alkali halide surfaces. Appl. Phys. A **78**, 837–841 (2004)
48. Hoffmann, R., Kantorovich, L.N., Baratoff, A., Hug, H.-J., Güntherodt, H.-J.: Sublattice identification in scanning force microscopy on alkali halide surfaces. Phys. Rev. Lett. **92**, 146103 (2004)
49. Burke, S.A., Mativetsky, J.M., Hoffmann, R., Grütter, P.: Nucleation and submonolayer growth of $C_{60}$ on KBr. Phys. Rev. Lett. **94**, 96102 (2005)
50. Schirmeisen, A., Weiner, D., Fuchs, H.: Single-atom contact mechanics: from atomic scale energy barrier to mechanical relaxation hysteresis. Phys. Rev. Lett. **97**(13), 136101 (2006)
51. Barth, C., Henry, C.R.: Imaging suzuki precipitates on NaCl: $Mg^2$ (001) by scanning force microscopy. Phys. Rev. Lett. **100**, 096101 (2008)
52. Ruschmeier, K., Schirmeisen, A., Hoffmann, R.: Atomic-scale force-vector fields. Phys. Rev. Lett. **101**(15), 1–4 (2008)
53. Cerda, M.A., Abad, J., Madgavkar, A., Martrou, D., Gauthier, S.: Step-induced tip polarity reversal investigated by dynamic force microscopy on KBr(001). Nanotechnology **19**, 045503 (2008)
54. Foster, A.S., Barth, C., Henry, C.R.: Chemical identification of ions in doped NaCl by scanning force microscopy. Phys. Rev. Lett. **102**, 256103 (2009)
55. Barth, C., Henry, C.R.: Surface double layer on (001) surfaces of alkali halide crystals: a scanning force microscopy study. Phys. Rev. Lett. **98**, 136804 (2007)
56. Egberts, P., Filleter, T., Bennewitz, R.: A kelvin probe force microscopy of charged indentation-induced dislocation structures in KBr. Nanotechnology **20**, 264005 (2009)
57. Hinaut, A.: Etude par Microscopie a Force Atomique en Mode Non Contact at par Microscopie a Sonde de Kelvin de Derives du Triphenylene sur KBr(001) dans l'ultra-vide. PhD thesis, Universite Toulouse III Paul Sabatier (2012)
58. Filleter, T., Maier, S., Bennewitz, R.: Atomic-scale yield and dislocation nucleation in KBr. Phys. Rev. B **73**(15), 1–10 (2006)
59. Pakarinen, O.H., Mativetsky, J.M., Gulans, A., Puska, M.J., Foster, A.S., Grütter, P.: Role of van der Waals forces in the adsorption and diffusion of organic molecules on an insulating surface. Phys. Rev. B **80**, 85401 (2009)
60. Such, B., Trevethan, T., Glatzel, T., Kawai, S., Zimmerli, L., Meyer, E., Shluger, A.L., Amijs, C.H.M., de Mendoza, P., Echavarren, A.M.: Functionalized truxenes: adsorption and diffusion of single molecules on the KBr(001) surface. ACS Nano **4**, 3429–3439 (2010)
61. Nony, L., Gnecco, E., Baratoff, A., Alkauskas, A., Bennewitz, R., Pfeiffer, O., Maier, S., Wetzel, A., Meyer, E., Gerber, Ch.: Observation of individual molecules trapped on a nanostructured insulator. Nano Lett. **4**, 2185–2189 (2004)
62. Kunstmann, T., Schlarb, A., Fendrich, M., Wagner, T., Möller, R., Hoffmann, R.: Dynamic force microscopy study of 3,4,9,10-perylenetetracarboxylic dianhydride on KBr (001). Phys. Rev. B **71**, 121403 (2005)
63. Dienel, T., Loppacher, C., Mannsfeld, S.C.B., Forker, R., Fritz, T.: Growth-mode-induced narrowing of optical spectra of an organic adlayer. Adv. Mater. **20**, 959–963 (2008)

64. Burke, S.A., Ji, W., Mativetsky, J.M., Topple, J.M., Fostner, S., Gao, H.-J., Guo, H., Grütter, P.: Strain induced dewetting of a molecular system: bimodal growth of PTCDA on NaCl. Phys. Rev. Lett. **100**, 186104 (2008)
65. Burke, S.A., LeDue, J.M., Topple, J.M., Fostner, Sh, Grütter, P.: Relating the functional properties of an organic semiconductor to molecular structure by nc-AFM. Adv. Mater. **21** (20), 2029–2033 (2009)
66. Hinaut, A., Pujol, A., Chaumeton, F., Martrou, D., Gourdon, A., Gauthier, S.: An NC-AFM and KPFM study of the adsorption of a triphenylene derivative on KBr(001). Beilstein J. Nanotechnol. **3**, 221–229 (2012)
67. Nony, L., Bocquet, F., Para, F., Chérioux, F., Duverger, E., Palmino, F., Luzet, V., Loppacher, C.: Dipole-driven self-organization of zwitterionic molecules on alkali halide surfaces. Beilstein J. Nanotechnol. **3**, 285–293 (2012)
68. Bocquet, F., Nony, L., Mannsfeld, S.C.B., Oison, V., Pawlak, R., Porte, L., Loppacher, C.: Inhomogeneous relaxation of a molecular layer on an insulator due to compressive stress. Phys. Rev. Lett. **108**(20), 206103 (2012)
69. Barth, C., Gingras, M., Foster, A.S., Gulans, A., Félix, G., Hynninen, T., Peresutti, R., Henry, C.R.: Two-dimensional nanostructured growth of nanoclusters and molecules on insulating suzuki surfaces. Adv. Mater. **24**, 3228–3232 (2012)
70. Hoff, B., Gingras, A., Foster, S., Peresutti, R., Henry, C.R., Barth, C.: Mechanisms of the adsorption and self-assembly of molecules with polarized functional groups on insulating surfaces. J. Phys. Chem. C **118**, 14569–14578 (2014)
71. Barth, C., Henry, C.R.: High-resolution imaging of gold clusters on KBr(001) surfaces investigated by dynamic scanning force microscopy. Nanotechnology **15**, 1264–1272 (2004)
72. Pakarinen, O.H., Barth, C., Foster, A.S., Henry, C.R.: High-resolution scanning force microscopy of gold nanoclusters on the KBr (001) surface. Phys. Rev. B **73**, 235428 (2006)
73. Barth, C., Henry, C.R.: Gold nanoclusters on alkali halide surfaces: charging and tunneling. Appl. Phys. Lett. **89**, 252119 (2006)
74. Mativetsky, J.M., Burke, S.A., Fostner, S., Grutter, P.: Nanoscale pits as templates for building a molecular device. Small **3**, 818–821 (2007)
75. Mativetsky, J.M., Fostner, S., Burke, S.A., Grutter, P.: High-resolution investigation of metal nanoparticle growth on an insulating surface. Phys. Rev. B **80**, 45430 (2009)
76. Glatzel, T., Zimmerli, L., Koch, S., Kawai, S., Meyer, E.: Molecular assemblies grown between metallic contacts on insulating surfaces. Appl. Phys. Lett. **94**, 063303 (2009)
77. Barth, C., Claeys, C., Henry, C.R.: Surface preparation of hard ionic crystals by ultrahigh vacuum cleavage. Rev. Sci. Instr. **76**, 083907 (2005)
78. Barth, C., Henry, C.R.: Kelvin probe force microscopy on surfaces of UHV cleaved ionic crystals. Nanotechnology **17**, S155–S161 (2006)
79. Cabailh, G., Henry, C.R., Barth, C.: Thin NaCl films on silver (001): Island growth and work function. New J. Phys. **14**, 103037 (2012)
80. Zhang, Y., Cattrall, R.W., McKelvie, I.D., Kolev, S.D.: Gold, an alternative to platinum group metals in automobile catalytic converters. Gold Bull. **44**, 145–153 (2011)
81. Foster, A.S., Trevethan, T., Shluger, A.L.: Structure and diffusion of intrinsic defects, adsorbed hydrogen, and water molecules at the surface of alkali-earth fluorides calculated using density functional theory. Phys. Rev. B **80**, 115421 (2009)
82. Hofer, W.A., Foster, A.S., Shluger, A.L.: Theories of scanning probe microscopes at the atomic scale. Rev. Mod. Phys. **75**, 1287–1331 (2003)
83. McKenna, K., Trevethan, T., Shluger, A.L.: Interplay between adsorbate diffusion and electron tunneling at an insulating surface. Phys. Rev. B **82**, 85427 (2010)
84. Frenkel, Y.: Kinetic Theory of Liquids. The Clarendon Press, Oxford (1946)
85. Whitworth, R.W.: Charged dislocations in ionic crystals. Adv. Phys. **24**, 203–304 (1975)
86. Kliewer, K.L., Koehler, J.S.: Space charge in ionic crystals. I. General approach with application to NaCl. Phys. Rev. **140**, A1226–A1240 (1965)
87. Kliewer, K.L.: Space charge in ionic crystals. II. The electron affinity and impurity accumulation. Phys. Rev. **140**, A1241–A1246 (1965)

88. Kliewer, K.: Space charge in ionic crystals—III. Silver halides containing divalent cations. J. Phys. Chem. Solids **27**, 705–717 (1966)
89. Poeppel, R.B., Blakely, J.M.: Origin of equilibrium space charge potentials in ionic crystals. Surf. Sci. **15**, 507–523 (1969)
90. Butman, M.F., Smirnov, A.A., Kudin, L.S., Munir, Z.A.: Determination of the sign of the intrinsic surface charge in alkali halides from ionic sublimation measurements. Surf. Sci. **458**, 106–112 (2000)
91. Baetzold, R.C.: Computation of the energetics of surface vacancy and interstitial generation in silver halide. Phys. Rev. B **52**, 11424–11431 (1995)
92. Blakely, J.M., Danyluk, S.: Space charge regions at silver halide surfaces: effects of divalent impurities and halogen pressure. Surf. Sci. **40**(1), 37–60 (1973)
93. Rao, A., Gnecco, E., Marchetto, D., Mougin, K., Schönenberger, M., Valeri, S., Meyer, E.: The analytical relations between particles and probe trajectories in atomic force microscope nanomanipulation. Nanotechnology **20**(11), 115706 (2009)
94. Barth, C., Foster, A.S., Reichling, M., Shluger, A.L.: Contrast formation in atomic resolution scanning force microscopy on $CaF_2$ (111): experiment and theory. J. Phys.: Condens. Matter **13**, 2061 (2001)
95. Barth, C., Pakarinen, O.H., Foster, A.S., Henry, C.R.: Imaging nanoclusters in the constant height mode of the dynamic SFM. Nanotechnology **17**, S128–S136 (2006)
96. Pakarinen, O.H., Barth, C., Foster, A.S., Henry, C.R.: Imaging the real shape of nanoclusters in scanning force microscopy. J. Appl. Phys. **103**, 054313 (2008)
97. Harding, J.H., Stoneham, A.M., Venables, J.A.: Predicting nucleation and growth processes: atomistic modeling of metal atoms on ionic substrates. Phys. Rev. B **57**(11), 6715–6719 (1998)
98. Hakala, M.H., Pakarinen, O.H., Foster, A.S.: First-principles study of adsorption, diffusion, and charge stability of metal adatoms on alkali halide surfaces. Phys. Rev. B **78**(4), 045418 (2008)
99. Omicron NanoTechnology GmbH, Taunusstein (Germany)
100. Nanosurf AG, Liestal (Switzerland)
101. Eigler, D.M., Lutz, C.P., Rudge, W.E.: Atomic switch realised with the scanning tunnelling microscope. Nature **352**, 600–603 (1991)
102. Sweetman, A., Jarvis, S., Danza, R., Moriarty, P.: Effect of the tip state during qPlus noncontact atomic force microscopy of Si (100) at 5 K: probing the probe. Beilstein J. Nanotechnol. **3**(1), 25–32 (2012)
103. Sweetman, A., Stannard, A., Sugimoto, Y., Abe, M., Morita, S., Moriarty, P.: Simultaneous noncontact AFM and STM of Ag: Si(111)-($\sqrt{3} \times \sqrt{3}$)R30$^0$. Phys. Rev. B **87**(7), 075310 (2013)
104. Yurtsever, A., Sugimoto, Y., Tanaka, H., Abe, M., Morita, S., Ondráček, M., Pou, P., Pérez, R., Jelínek, P.: Force mapping on a partially H-covered Si(111)-(7 × 7) surface: influence of tip and surface reactivity. Phys. Rev. B: Condens. Matter **87**(15), 155403 (2013)
105. Baykara, M.Z., Schwendemann, T.C., Altman, E.I., Schwarz, U.D.: Three-dimensional atomic force microscopy—taking surface imaging to the next level. Adv. Mater. **22**, 2838–2853 (2010)
106. Rahe, Ph, Schutte, J., Schniederberend, W., Reichling, M., Abe, M., Sugimoto, Y., Kuhnle, A.: Flexible drift-compensation system for precise 3D force mapping in severe drift environments. Rev. Sci. Instr. **82**(6), 063704 (2011)
107. Bennewitz, R., Schar, S., Barwich, V., Pfeiffer, O., Meyer, E., Krok, F., Such, B., Kolodzej, J., Szymonski, M.: Atomic-resolution images of radiation damage in KBr. Surf. Sci. **474**, L197–L202 (2001)
108. Barth, C., Henry, C.R.: NaCl(001) surfaces nanostructured by Suzuki precipitates: a scanning force microscopy study. New J. Phys. **11**, 043003 (2009)
109. Pacchioni, G.: Theory of point defects at the MgO surface. In: Woodruff, D.P. (ed.) The Chemical Physics of Solid Surfaces. Elsevier, Amsterdam (2002)

110. Diebold, U.: The surface science of titanium dioxide. Surf. Sci. Rep. **48**, 53 (2003)
111. Pang, C.L., Lindsay, R., Thornton, G.: Chemical reactions on rutile $TiO_2$(110). Chem. Soc. Rev. **37**(10), 2328–2353 (2008)
112. Dohnálek, Z., Lyubinetsky, I., Rousseau, R.: Thermally-driven processes on rutile $TiO_2$(110)-(1 × 1): a direct view at the atomic scale. Prog. Surf. Sci. **85**(5), 161–205 (2010)
113. Paier, J., Penschke, Ch., Sauer, J.: Oxygen defects and surface chemistry of ceria: quantum chemical studies compared to experiment. Chem. Rev. **113**(6), 3949–3985 (2013)
114. Kelber, J.A.: Alumina surfaces and interfaces under non-ultrahigh vacuum conditions. Surf. Sci. Rep. **62**(7), 271–303 (2007)
115. Freund, H.J., Pacchioni, G.: Oxide ultra-thin films on metals: new materials for the design of supported metal catalysts. Chem. Soc. Rev. **37**, 2224–2242 (2008)
116. Chen, M.S., Goodman, D.W.: Ultrathin, ordered oxide films on metal surfaces. J. Phys.: Condens. Matter **20**, 264013 (2008)
117. Nilius, N.: Properties of oxide thin films and their adsorption behavior studied by scanning tunneling microscopy and conductance spectroscopy. Surf. Sci. Rep. **64**, 595–659 (2009)
118. Artiglia, L., Cavaliere, E., Vascon, A., Bondino, F., Rizzi, G.A., Gavioli, L., Granozzi, G.: Reactivity of Fe nanoparticles on $TiO_x$/Pt(111): a complete surface science investigation. J. Phys. Chem. C **115**, 15812–15821 (2011)
119. Pacchioni, G., Freund, H.-J.: Electron transfer at oxide surfaces. The MgO paradigm: from defects to ultrathin films. Chem. Rev. **113**, 4035–4072 (2013)
120. Lauritsen, J.V., Reichling, M.: Atomic resolution non-contact atomic force microscopy of clean metal oxide surfaces. J. Phys.: Condens. Matter **22**, 263001 (2010)
121. Palacios-Lidon, E., Henry, C.R., Barth, C.: Kelvin probe force microscopy in surface chemistry: reactivity of Pd nanoparticles on highly oriented pirolytic graphite. ACS Catal. **4** (6), 1838–1844 (2014)

# Imaging of Defects on Ge(001):H by Non-contact Atomic Force Microscopy

**Bartosz Such, Marek Kolmer, Szymon Godlewski, Jakub Lis, Janusz Budzioch, Mateusz Wojtaszek and Marek Szymonski**

**Abstract** The structure of a hydrogenated Ge(001) surface is examined by means of non-contact atomic force microscopy (NC-AFM) at 5 K. Three-dimensional force spectroscopy of a bond defect surrounded by a passivated area shows a qualitatively distinct spatial dependence of the interaction. The force curve over a defect has a range where the short-range attractive force is detected, whereas over the hydrogen-terminated bonds, it is missing: atomic-scale imaging of hydrogens is possible only in the repulsive range. Thus, NC-AFM allows for clear determination of the different chemical activities of the defect compared to the surrounding passivated area.

**Keywords** NC-AFM · Ge(001) · Surface defect

## 1 Introduction

New concepts in the fabrication of atomic-scale wires have inspired renewed interest in hydrogenated surfaces of germanium and silicon. It is proposed that the selective desorption of hydrogen from the passivated semiconductor surface could provide the means for efficient creation of surface-supported chains of dangling bond wires [1–8]. Upon passivation by hydrogen, each dangling bond present on a dimerized $c(4 \times 2)/p(2 \times 2)$ surface of Ge(001) binds a hydrogen atom [9, 10]. That removes the tilt of dimers and creates a $2 \times 1$ surface structure. However, the hydrogenation process usually is not complete, and a certain population of non-passivated dangling bonds and other defects is still present on the surface. Such defects are an exciting testing ground for scanning probe techniques, since they

B. Such (✉) · M. Kolmer · S. Godlewski · J. Lis · J. Budzioch ·
M. Wojtaszek · M. Szymonski
Centre for Nanometer-Scale Science and Advanced Materials (NANOSAM),
Faculty of Physics, Astronomy and Applied Computer Science,
Jagiellonian University, Reymonta 4, 30-059 Krakow, Poland
e-mail: bartosz.such@uj.edu.pl



P. Moriarty and S. Gauthier (eds.), *Imaging and Manipulation of Adsorbates Using Dynamic Force Microscopy*, Advances in Atom and Single Molecule Machines,
DOI 10.1007/978-3-319-17401-3_7

form chemically active centers protruding from the surface and surrounded by a passivated, chemically inert area.

In this chapter, we report the results of a non-contact atomic force microscopy (NC-AFM) study of such defects performed at cryogenic temperatures. NC-AFM at low temperatures has provided quite a number of interesting results recently, mainly thanks to the introduction of tuning forks as sensors [11, 12]. A variety of systems were investigated such as metals [13, 14], HOPG [15], Si(001) [16], nanodiamonds [17], and insulators [18], as well as molecular [19–23] and metallic [24] adsorbates. While NC-AFM in general has been proven to be able to distinguish different species on a surface [25] at the atomic scale, the aforementioned studies were not concerned with examining radically different chemical activities on a single sample. This is also true for investigations of surfaces of hydrogenated semiconductor surfaces [26–30]. Local chemical activity at the atomic scale, however, is of considerable interest in many applications. This particularly true for studies of single organic molecules, where adsorption of a CO molecule on a tip is typically used to provide the chemical inactivity essential for the outcome (i.e., resolving the internal structure of a molecule) [19, 20, 22] of those experiments.

## 2 Experimental

The experiments were carried out in a UHV system equipped with a LT-STM/AFM instrument (Omicron) operating at 5 K. Commercially available tuning fork-based sensors (qPlus sensors) are equipped with chemically etched tungsten tips. Since simultaneous STM and NC-AFM measurements with tuning fork-based systems are often marred by problems related to a cross talk between the measurement channels, our experiment is performed with tips which are non-conductive, probably due to a thick oxide layer on the apex. Lack of the tunneling current measurement is the price for reliable measurement of the frequency shift of the sensor and for avoiding artifacts in the dissipation channel. The Ge(001) samples were prepared by repeated cycles of sputtering ($Ar^+$, 600 eV, 750 °C) until a clean $c(4 \times 2)/p(2 \times 2)$ surface was obtained, as checked by low energy electron diffraction (LEED) and LT-STM measurements. The substrate was then exposed to atomic hydrogen dosed with the use of a homebuilt cracker ($1 \times 10^{-7}$ mbar of $H_2$ partial pressure, samples kept at 200 °C). Following this process, a well-hydrogenated surface with a limited number of defects is observed.

## 3 Results

An NC-AFM image of a hydrogenated Ge(001) surface is shown in Fig. 1. The main features in the constant frequency shift image are double rows of depressions, i.e., positions where the tip had to approach closer to the surface in order to

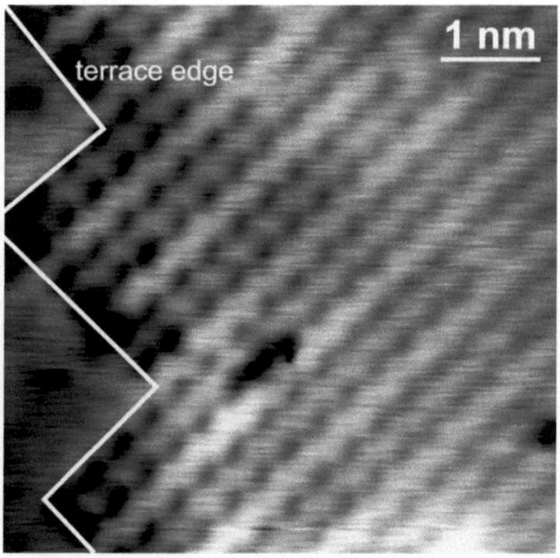


**Fig. 1** NC-AFM constant frequency shift image of a hydrogenated Ge(001) surface; $f_0$ = 22,500 Hz, $\Delta f = -10$ Hz, $A$ = 500 pm, and $Q$ = 10,000

maintain the required frequency shift. The rows on adjacent terraces separated by a monolayer step are perpendicular to each other, as expected for a reconstructed Ge (001) surface. The separation between the protrusions in a single row is about 0.37 nm, while the double rows are separated by 0.82 nm. Those values correspond quite closely to the expected distances between hydrogen atoms adsorbed on a Ge dimer and the distance between dimer rows on the Ge(001) surface (0.39 nm [31] and $2 * a_0\sqrt{2} = 0.8$ nm, respectively, where $a_0$ is the bulk lattice constant of Ge). Therefore, it is plausible that the depressions can be identified as hydrogen atoms bound on the surface dimers. Occasionally, defects can be observed on the surface, located over one of the hydrogens in a dimer row. In the atomically resolved images, they are depicted as even deeper depressions than the hydrogen atoms.

In order to gain more insight into the contrast formation mechanism, three-dimensional (3D) frequency shift-distance spectroscopy was performed. A series of constant-height images (32 in total) was taken over a small ($1.8 \times 1.8$ nm$^2$) area with decreasing tip-surface separation. In order to elucidate the details of the tip interaction with species present on the surface, an area containing both hydrogen-passivated dimers and a defect was chosen. A representative image taken from the series is shown in Fig. 2. The brighter colors depict increasingly negative frequency shifts, i.e., increasingly attractive interactions. A series of such images forms a 3D field of frequency shift experienced by a sensor over the surface. A frequency shift versus distance curve over a site extracted from the 3D dataset is presented in Fig. 2b, together with a force versus distance curve calculated according to the procedure introduced by Sader and Jarvis [32]. The force curve shows significant interaction as far as 2 nm from the surface. Therefore, the curve is dominated by long-range interactions, perhaps arising from a relatively large radius of curvature of the tip. The long-range forces are not site specific, and hence, the curves taken over different sites on the surface are virtually unrecognizable at this scale. As a result, in order to focus only on the short range, chemically specific forces in the

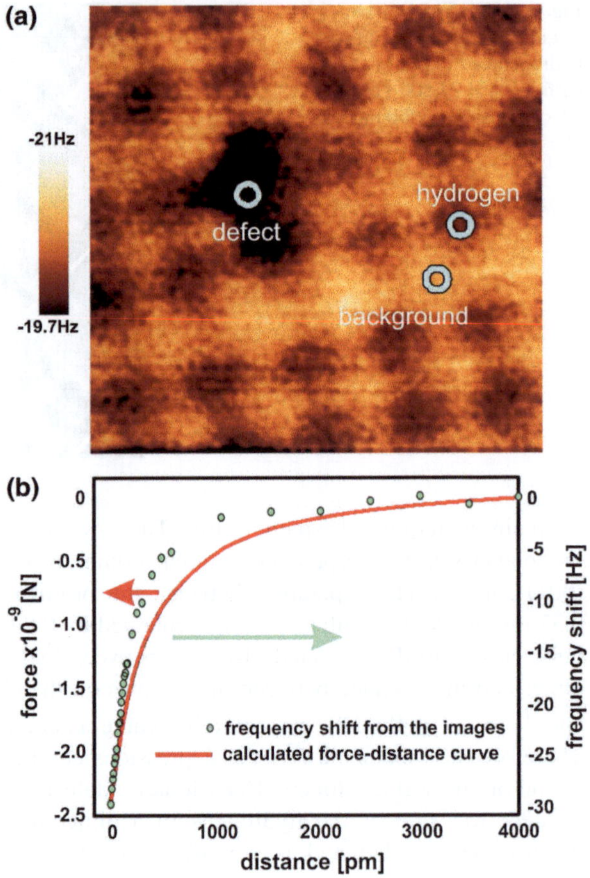


**Fig. 2** 3D spectroscopy of a Ge(001):H surface: **a** one of the constant-height images forming 3D dataset with marked 'background,' 'defect,' and 'hydrogen' positions; **b** frequency shift versus distance dependence (*circles, scale on the right axis*) over an arbitrary point within the scan area taken from the consecutive images and the force versus distance curve calculated from the frequency shift data (*line, scale on the left axis*)

following, only *differences* between the forces over the characteristic sites are analyzed.

The position between the dimer rows does not show any contrast evolution upon decreasing the tip-surface distance until a severe tip relaxation occurs. Therefore, it is likely that the influence of a short-range component is small at that position and the contrast evolution over hydrogen atoms or the defect can be analyzed with respect to that site (marked in Fig. 2a as 'background'). The curves presented in Fig. 3 show the dependence of the difference between the force over the 'defect' site and the 'background' site $F_D$ (red line), and between the 'hydrogen' site and the 'background' site $F_H$ (blue line), on the distance. The position '$z = 0$' is chosen arbitrarily as the position of the closest image taken. For large tip-surface distances, the lines are virtually unrecognizable, meaning that there is no contrast visible in the constant-height force image. The first discrepancies between the curves start around a distance of 300 pm, where the curve $F_D$ becomes negative and reaches a magnitude of almost 10 pN. This means that the tip is attracted to the defect in

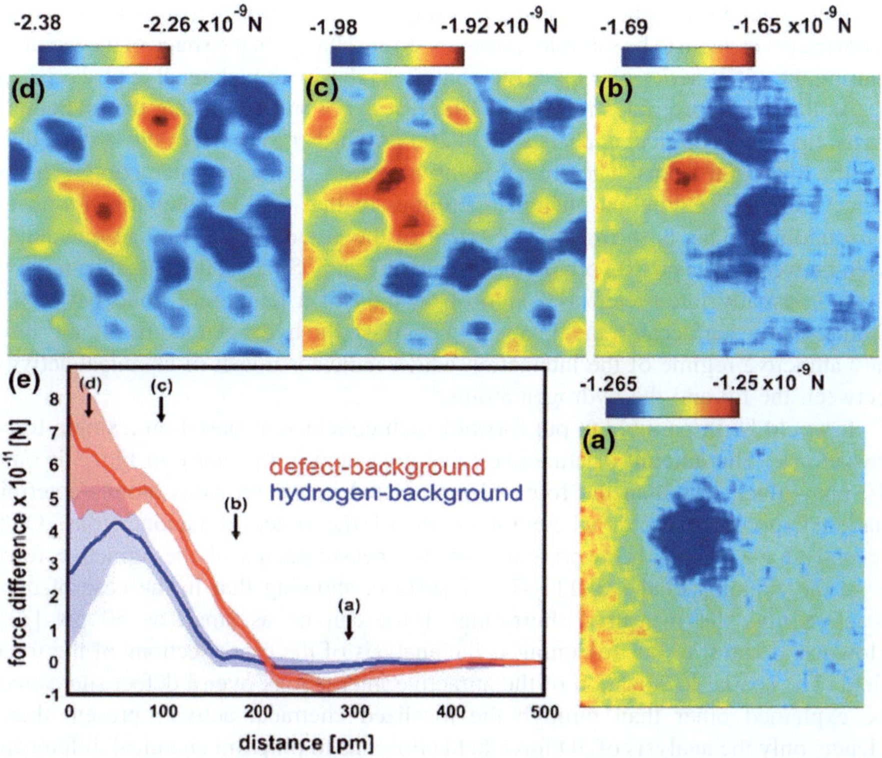


**Fig. 3** Evolution of the force over a hydrogen atom (*blue line*) and a defect (*red line*) with respect to the 'background' site; the *shaded areas* represent the standard deviation of the force difference calculated for the curves extracted for the areas of interest from the smoothed 3D force field (**a–d**) and the cross sections of a force field taken at characteristic tip-surface distances marked on the plot of force difference versus distance (**e**)

comparison with the background. We note that this value of the force difference is close to our detection limit. However, the region of increased attraction located over the area of the defect is clearly visible in Fig. 3a, supporting our interpretation. The curve $F_H$ is still close to zero in that region. The constant-height cross section of the 3D force field corresponding to that region is shown in Fig. 3a, showing the area of attraction over the defect with no apparent contrast around. While the tip approaches the surface, the curve $F_D$ changes its sign and increases rapidly. The tip starts to experience the repulsion over the defect, and the force difference climbs to $5 \times 10^{-11}$ N. The $F_H$ curve is still close to zero. The cross section of the force field, presented in Fig. 3b, shows a small area of relative repulsion located over the defect site and hardly any contrast around.

When the tip approaches further, the curve $F_H$ rises to positive values indicating that a repulsive interaction appears over the hydrogen atoms. Curve $F_D$ is saturated. Figure 3c shows clear contrast on the hydrogenated surface, while the defect loses its circular symmetry. Figure 3d, corresponding to the position where both force difference curves seem to be saturated, shows a strong change in the shape of the features. During the entire series of images, no contrast in dissipation channel is observed.

As the defect protrudes from the surface, it is not surprising that the interaction of the tip with the defect appears around 200 pm further away from the surface than for the hydrogen atoms. The crucial point, however, is that the interaction over the defect indicates a different chemical activities of that site, in comparison with its surroundings. This is shown by the extent of the attractive short-range interaction seen between 200 and 300 pm on the $z$-scale of Fig. 3. Such a site-specific attractive force is usually interpreted as a sign of creation of a temporary bond between a feature on the surface and the tip apex. In contrast, the hydrogen atoms do not show any attractive regime of the interaction which indicates a lack of chemical activity between the tip and the hydrogen atoms.

It would be impossible to put forward such conclusions based on a single force curve only. The amount of attractive force presented in the curve in Fig. 3 is only 10 pN—much less than the forces demonstrated in the previous atom—determination experiments [25], where forces are of the order of nanonewtons. Only recently have reports been presented of the measurements of the attractive force over the adatoms of the Si(111)-7 × 7 surface, showing that in the case of non-reactive tips, the measured short-range force can be as small as 50 pN [33]. However, central to our reasoning is the analysis of the cross sections of the force field. The spatial localization of the attractive interactions over a defect site cannot be explained other than through the localized chemical activity present there. Hence, only the analysis of 3D force field allows us to pinpoint chemical differences between the various sites on the surface.

The small value of the measured attractive force is certainly connected to the atomic structure of the tip. The exact structure of the tip is almost always a problem in NC-AFM experiments. In our work, the tip apex is formed by deliberate contacts with the surface. We note that it is quite difficult, and therefore rare, to obtain a tip which is stable while scanning over both the hydrogenated surface and the defects. This is not surprising since the requirements for the tip to image chemical species as distinct as the dangling bond and the passivated surface are quite demanding, and the tip structure is perhaps quite peculiar. The small value of the attractive force over the defect presented in Fig. 3 suggests that most likely the tip is not terminated by clean germanium, but it is partially passivated with hydrogen. A more exact termination of the tip apex, however, could only be corroborated by extensive DFT calculations.

In summary, we have presented atomically resolved images and 3D force spectroscopy of the hydrogenated Ge(001) surface. NC-AFM spectroscopy revealed different local chemical activity of a defect, most likely a dangling bond, with respect to the passivated Ge surface in the vicinity of the defect. The difference is marked by the 100 pm long span of the attractive short-range interaction between the tip and the defect. In contrast, hydrogen-passivated bonds do not exhibit any chemical activity and there is no site-specific interaction between the hydrogen and the tip until the tip apex enters the range of the overlapping electronic states. Thus, we show that NC-AFM is capable of imaging and differentiating, at the atomic scale, species of contrasting chemical properties, and when corroborated by DFT calculations in future, it could bring even quantitative measurement of the local chemical activity.

**Acknowledgments** Founding for this research has been provided by European Community under Large-scale Integrating Project in FET Proactive of the 7th FP entitled 'Atomic scale and single molecule logic gate technologies, AtMol.' The research was carried out with the equipment purchased thanks to the financial support of the European Regional Development Fund in the framework of the Polish Innovation Economy Operational Program (contract no. POIG.02.01.00-12-023/08). S.G. would like to acknowledge financial support received from the Foundation for Polish Science within START program (2010 and 2011).

## References

1. Lyding, J.W., Shen, T.C., Hubacek, J.S., Tucker, J.R., Abeln, G.C.: Nanoscale patterning and oxidation of H-passivated Si(100)-2X1 surfaces with an ultrahigh-vaccum scanning tunneling microscope. Appl. Phys. Lett. **64**, 2010 (1994)
2. Ample, F., Duchemin, I., Hliwa, M., Joachim, C.: Single OR molecule and OR atomic circuit logic gates interconnected on a Si(100)H surface. J. Phys. Cond. Mat. **23**, 125303 (2011)
3. Kawai, H., Yeo, Y.K., Saeys, M., Joachim, C.: Conductance decay of a surface hydrogen tunneling junction fabricated along a Si(001)-(2x1)-H atomic wire. Phys. Rev. B **81**, 195316 (2010)
4. Scappucci, G., Capellini, G., Lee, W.C.T., Simmons, M.Y.: Atomic-scale patterning of hydrogen terminated Ge(001) by scanning tunneling microscopy. Nanotechnology **20**, 495302 (2009)
5. Shen, T.C., Wang, C., Abeln, G.C., Tucker, J.R., Lyding, J.W., Avouris, P., Walkup, R.E.: Atomic-scale desorption through electronic and vibrational-excitation mechanisms. Science **268**, 1590 (1995)
6. Soukiassian, L., Mayne, A.J., Carbone, M., Dujardin, G.: Atomic wire fabrication by STM induced hydrogen desorption. Surf. Sci. **528**, 121 (2003)
7. Kolmer, M., Godlewski, S., Kawai, H., Such, B., Krok, F., Saeys, M., Joachim, C., Szymonski, M.: Electronic properties of STM-constructed dangling-bond dimer lines on a Ge (001)-(2x1):H surface. Phys. Rev. B. **86**, 125307 (2012)
8. Kolmer, M., Godlewski, S., Zuzak, R., Wojtaszek, M., Rauer, C., Thuaire, A., Hartmann, J.-M., Moriceau, H., Joachim, C., Szymonski, M.: Atomic scale fabrication of dangling bond structures on hydrogen passiyated Si(001) wafers processed and nanopackaged in a clean room environment. Appl. Surf. Sci. **288**, 83 (2014)
9. Lee, J.Y., Jung, S.J., Maeng, J.Y., Cho, Y.E., Kim, S., Jo, S.K.: Atomic-scale structural evolution of Ge(100) surfaces etched by H and D. Appl. Phys. Lett. **84**, 5028 (2004)
10. Maeng, J.Y., Lee, J.Y., Cho, Y.E., Kim, S., Jo, S.K.: Surface dihydrides on Ge(100): A scanning tunneling microscopy study. Appl. Phys. Lett. **81**, 3555 (2002)
11. Giessibl, F.J.: High-speed force sensor for force microscopy and profilometry utilizing a quartz tuning fork. Appl. Phys. Lett. **73**, 3956 (1998)
12. Giessibl, F.J.: Atomic resolution on Si(111)-(7x7) by noncontact atomic force microscopy with a force sensor based on a quartz tuning fork. Appl. Phys. Lett. **76**, 1470 (2000)
13. König, T., Simon, G.H., Rust, H.P., Heyde, M.: Atomic resolution on a metal single crystal with dynamic force microscopy. Appl. Phys. Lett. **95**, 083116 (2009)
14. Such, B., Glatzel, T., Kawai, S., Meyer, E., Turansky, R., Brndiar, J., Stich, I.: Interplay of the tip-sample junction stability and image contrast reversal on a Cu(111) surface revealed by the 3D force field. Nanotechnology **23**, 045705 (2012)

15. Albers, B.J., Schwendemann, T.C., Baykara, M.Z., Pilet, N., Liebmann, M., Altman, E.I., Schwarz, U.D.: Three-dimensional imaging of short-range chemical forces with picometre resolution. Nat. Nanotechnol. **4**, 307 (2009)
16. Sweetman, A., Jarvis, S., Danza, R., Bamidele, J., Gangopadhyay, S., Shaw, G.A., Kantorovich, L., Moriarty, P.: Toggling bistable atoms via mechanical switching of bond angle. Phys. Rev. Lett. **106**, 136101 (2011)
17. Pawlak, R., Glatzel, Th, Pichot, V., Schmidlin, L., Kawai, S., Fremy, S., Spitzer, D., Meyer, E.: Local detection of nitrogen-vacancy centers in a nanodiamond monolayer. Nano Lett. **13**, 5803 (2013)
18. Such, B., Glatzel, T., Kawai, S., Koch, S., Meyer, E.: Three-dimensional force spectroscopy of KBr(001) by tuning fork-based cryogenic noncontact atomic force microscopy. J. Vac. Sci. Tech. B **28**, C4B1 (2010)
19. Gross, L., Mohn, F., Moll, N., Liljeroth, P., Meyer, G.: The chemical structure of a molecule resolved by atomic force microscopy. Science **325**, 1110 (2009)
20. Gross, L., Mohn, F., Moll, N., Meyer, G., Ebel, R., Abdel-Mageed, W.M., Jaspars, M.: Organic structure determination using atomic-resolution scanning probe microscopy. Nat. Chem. **2**, 821 (2010)
21. Pawlak, R., Kawai, S., Fremy, S., Glatzel, T., Meyer, E.: Atomic-scale mechanical properties of orientated C(60) molecules revealed by noncontact atomic force microscopy. ACS Nano **5**, 6349 (2011)
22. Kawai, S., Sadeghi, A., Feng, X., Lifen, P., Pawlak, R., Glatzel, Th., Willand, A., Orita, A., Otera, J., Goedecker, S., Meyer, E.: Obtaining detailed structural information about supramolecular systems on surfaces by combining high-resolution force microscopy with ab initio calculations. ACS Nano **7**, 9098 (2013)
23. Kawai, S., Koch, M., Gnecco, E., Sadeghi, A., Pawlak, R., Glatzel, Th., Schwarz, J., Goedecker, S., Hecht, S., Baratoff, A., Grill, L., Meyer, E.: Quantifying the atomic-level mechanics of single long physisorbed molecular chains. Proc. Nat. Am. Sci. **111**, 3968 (2014)
24. Ternes, M., Lutz, C.P., Hirjibehedin, C.F., Giessibl, F.J., Heinrich, A.J.: The force needed to move an atom on a surface. Science **319**, 1066 (2008)
25. Sugimoto, Y., Pou, P., Abe, M., Jelinek, P., Perez, R., Morita, S., Custance, O.: Chemical identification of individual surface atoms by atomic force microscopy. Nature **446**, 64 (2007)
26. Araragi, S., Yoshimoto, A., Nakata, N., Sugawara, Y., Morita, S.: Atomic resolution imaging of Si(100) 1 x 1 : 2H dihydride surface with noncontact atomic force microscopy (NC-AFM). Appl. Surf. Sci. **188**, 272 (2002)
27. Miura, N., Tsukada, M.: Theoretical analysis of tip effect on noncontact atomic force microscopy image of Si(100) 2 x 1 : H surface. Jpn. J. Appl. Phys. Part **1**(41), 306 (2002)
28. Morita, S., Sugawara, Y.: Atomically resolved imaging of Si(100)2 x 1, 2 x 1 : H and 1 x 1 : 2H surfaces with noncontact atomic force microscopy. Jpn. J. Appl. Phys. Part 1 **41**, 4857 (2002)
29. Nimmrich, M., Kittelmann, M., Rahe, P., Harneit, W., Mayne, A.J., Dujardin, G., Kuhnle, A.: Influence of charge transfer doping on the morphologies of C(60) islands on hydrogenated diamond C(100)-(2 x 1). Phys. Rev. B **85**, 035420 (2012)
30. Sharp, P., Jarvis, S., Woolley, R., Sweetman, A., Kantorovich, L., Pakes, C., Moriaty, P.: Identifying passivated dynamic force microscopy tips on H:Si(100). Appl. Phys. Lett. **100**, 233120 (2012)
31. Pflanz, S., Buchtler, R., Moritz, W., Over, H.: Dimer bond geometry in D/Ge(100)-(2x1): A low-energy electron-diffraction structure analysis. Phys. Rev. B **54**, R8313 (1996)
32. Sader, J.E., Jarvis, S.P.: Accurate formulas for interaction force and energy in frequency modulation force spectroscopy. Appl. Phys. Lett. **84**, 1801 (2004)
33. Yurtsever, A., Sugimoto, Y., Tanaka, H., Abe, M., Morita, S., Ondráček, M., Pou, P., Pérez, R., Jelínek, P.: Force mapping on a partially H-covered Si(111)-(7x7) surface: Influence of tip and surface reactivity. Phys. Rev. B **87**, 155403 (2013)

# Adsorption Structures of Amino Acids on Calcite(104)

**Felix Kling, Markus Kittelmann and Angelika Kühnle**

**Abstract** Elucidating the interaction details of proteins with the most stable cleavage plane of calcite, namely calcite(104), is of great importance for understanding the physicochemical mechanisms behind biomineralisation. In this context, amino acids are generally believed to serve as suitable model molecules, as they constitute the basic building blocks of proteins. In this work, we present a non-contact atomic force microscopy (NC-AFM) investigation of the adsorption of five proteinogenic amino acids on calcite(104) under ultra-high vacuum (UHV) conditions. For studying the structures formed from comparatively large amino acids, enantiopure tryptophan, tyrosine and aspartic acid molecules are deposited onto the surface held at room temperature (RT). These results are compared to the structures observed when depositing the two smallest amino acids, namely glycine and alanine. Our results reveal strikingly similar island structures with a $(5 \times 1)$ superstructure for the class of large amino acids despite the rather different side chains. The chirality of the molecules is unambiguously identified by a characteristic angle that is formed with respect to the $[42\bar{1}]$ substrate direction. The structures observed for glycine and alanine, on the other hand, differ substantially from each other and also from the $(5 \times 1)$ pattern revealed for the large amino acid. Our study illustrates that identifying general adsorption principles is difficult even in the case of rather simple molecular building blocks.

**Keywords** Self-assembly · Biomineralisation · Non-contact atomic force microscopy · Surface chirality

F. Kling · M. Kittelmann · A. Kühnle (✉)
Institute of Physical Chemistry, Department of Chemistry,
Johannes Gutenberg-University Mainz, Duesbergweg 10-14,
55099 Mainz, Germany
e-mail: kuehnle@uni-mainz.de

 
P. Moriarty and S. Gauthier (eds.), *Imaging and Manipulation of Adsorbates Using Dynamic Force Microscopy*, Advances in Atom and Single Molecule Machines,
DOI 10.1007/978-3-319-17401-3_8

# 1 Introduction

Amino acids, one of the most fundamental components of life, appear to have evolved spontaneously under early earth conditions, as has been indicated by the Miller–Urey experiment [1]. Amino acids constitute the basic building blocks of proteins, which are nature's "molecular machines". For instance, proteins can act as catalysts, facilitating chemical reactions, or nanoscale transporter, directing specific molecules towards their site of use.

Proteins are substantially involved in biomineralisation, nature's way of creating organic–inorganic hybrid materials with outstanding properties [2]. An impressive number of biominerals with tailor-made properties can be found in nature such as hybrid materials composed of proteins and calcite or aragonite in the shells of marine organisms.

Amino acids are interesting molecules not only due to their biological relevance, but also because they represent a prominent example for point chirality. Therefore, upon adsorption onto a surface, the formation of global chiral structures is possible and observed frequently. Besides the chirality, amino acids exist in a protonated, deprotonated or even zwitterionic state, significantly altering the adsorption structure [3]. Additionally, since amino acids contain freely rotatable bonds, different conformations and adsorption positions are feasible. This allows for forming quite complex structures and different phases depending on temperature and coverage [4].

In this chapter, we will focus on the adsorption of five amino acids onto the natural cleavage plane of calcite, namely calcite(104). Many studies exist in literature, addressing the adsorption behaviour of various amino acids on metal surfaces in ultra-high vacuum (UHV) [4–9]. Surprisingly, no molecular-resolution measurements of amino acids adsorbed to an insulating surface in UHV have been reported in literature so far.

Calcite (Fig. 1a) is one of the best studied materials for biomineralisation [2]. It can be found, e.g., in mollusc shells. For understanding the principles behind biomineralisation, some work has focused on calcite crystallisation in the presence of amino acids. An important observation is that adding enantiopure amino acids of different chirality can result in chirally shaped calcite crystals. This constitutes a most elegant way of showing how macroscopic chirality evolves from molecular chirality [10, 11].

Calcite is known to be an excellent surface for molecular self-assembly due to the high surface energy as compared to other non-conductive materials [12]. Therefore, a number of molecules have been observed to self-assemble on calcite (104) into ordered structures ranging from molecular islands [13] to wire-like structures [14].

We selected the amino acids tryptophan, tyrosine and aspartic acid (Fig. 1b–d) for deposition as they represent rather large amino acids. These molecules were used in their enantiopure form in both chiralities. Since most literature deals with smaller amino acids, only little is known so far about the adsorption of tyrosine [15, 16], tryptophan [17, 18] and aspartic acid [19]. Moreover, all these studies are

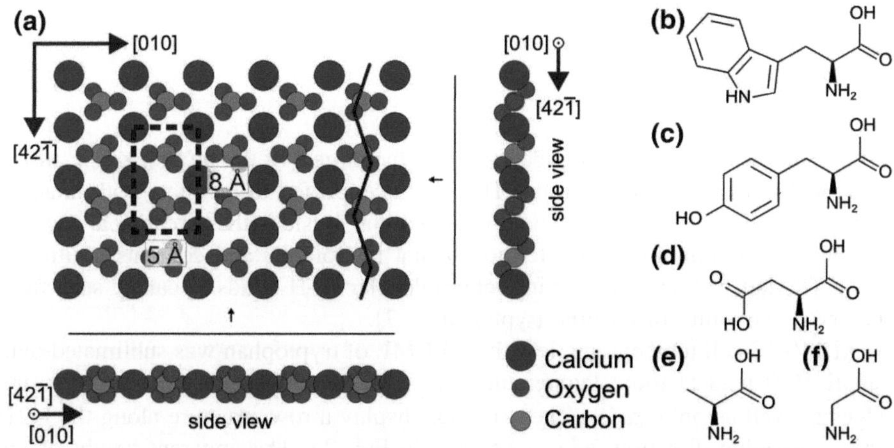


**Fig. 1** **a** Model of the calcite(104) surface and structural formulas of the amino acids used in this study, namely **b** L-tryptophan, **c** L-tyrosine and **d** L-aspartic acid as representatives for large amino acids and **e** L-alanine and **f** glycine constituting the smallest amino acids

limited to metallic surfaces. To compare the structures obtained for the three amino acids mentioned above, we also investigated the adsorption structures of the two smallest amino acids, namely glycine and alanine (Fig. 1e–f). Being the simplest amino acids, these two molecules have been studied extensively in the past, with a focus on understanding both structure formation [9, 20, 21] as well as adsorbate-induced surface reconstruction [22–24] on metallic surfaces.

## 2 Experimental Part

In this work, a variable-temperature atomic force microscope (VT-AFM from Omicron, Taunusstein, Germany) was used for performing non-contact atomic force microscopy (NC-AFM) imaging in the frequency modulated mode. All experiments were performed following a procedure as described previously [25]. The orientation of the calcite crystals was determined by its birefringence. All molecules were purchased from Sigma-Aldrich with purities larger than 98 %. For molecule deposition, a home-built Knudsen cell was used. The temperatures for sublimation were in the range of 110–145 °C for the larger amino acids. The corresponding deposition rates were less than 0.01 monolayers (ML) per minute. Sublimation times ranged from 30 min to 2 h. Alanine and glycine were sublimated at 80 °C for 5–30 min, the resulting coverage being in the order of 0.1 ML. Detail images have been corrected for thermal drift [26].

# 3 Results and Discussion

## 3.1 Tryptophan

Tryptophan was chosen mainly because it is the largest of the proteinogenic amino acids with a length of about 1 nm. This size is expected to facilitate both imaging and identification of the molecular conformation. Besides these technical aspects, the side chain contains an indole moiety that is aromatic and contains a nitrogen atom. The latter allows for forming intermolecular N–H bonds as can be seen from the crystal structure of racemic tryptophan [27].

At 110 °C cell temperature, less than 0.1 ML of tryptophan was sublimated onto calcite(104) held at room temperature (RT). Molecular islands form both at step edges as well as on terraces. These islands display a row structure along the $[42\bar{1}]$ crystallographic direction, as can be seen in Fig. 2a. The apparent height of the islands is about 4 Å, which can be tentatively assigned to flat-lying molecules, although height information in NC-AFM should be interpreted with care in general.

When zooming into the islands, the NC-AFM images reveal further details of the inner structure. As can be seen in Fig. 2b, c for D-tryptophan and L-tryptophan, respectively, the contrasts in the two high-resolution images are somewhat different,

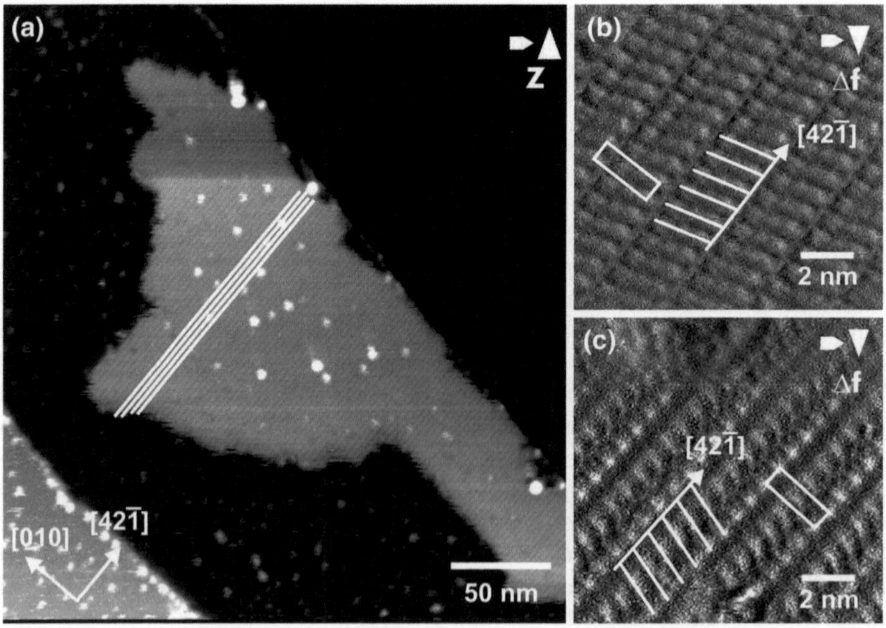


**Fig. 2** **a** Molecular island of D-tryptophan and high-resolution images of **b** D- and **c** L-tryptophan. The (5 × 1) surface cell is marked by a *rectangle*, the inner striped structure is indicated by *white lines*

which is ascribed to different tip terminations. However, both images clearly reveal that the rows along the $[42\bar{1}]$ direction are composed by a striped inner structure, resulting in a $(5 \times 1)$ unit cell. Interestingly, the stripes form a stacked pattern (marked by white lines in the figure) with an angle of 108° with respect to the $[42\bar{1}]$ direction. For L- and D-tryptophan, the same structure can be observed, but mirrored along the $[42\bar{1}]$ direction. This finding suggests that a distinction between L- and D-enantiomers is possible based on the chirality of the observed patterns.[1]

## 3.2 Tyrosine

Tyrosine is the second largest proteinogenic amino acid, featuring a phenol moiety in its side chain and, hence, containing one additional hydroxyl group. This group provides an additional possibility for the formation of hydrogen bonds. The crystal structure of L-tyrosine is known to be stabilised mainly by hydrogen bonds between the amino acid and hydroxyl groups [28].

A cell temperature of 115 °C proved to be optimum for sublimation of tyrosine. Upon deposition of less than 0.1 ML of tyrosine onto calcite(104) held at RT, islands form similar to the situation revealed after deposition of tryptophan (Fig. 3a). The islands again exhibit an apparent height of 4 Å, indicating flat-lying molecules.

High-resolution images again display a $(5 \times 1)$ superstructure (Fig. 3 b, c) and allow for distinguishing between the two enantiomers. As for tryptophan, an angle of 108° is formed between the molecular stripes and the $[42\bar{1}]$-direction. This observation suggests that the change in side chain does not significantly influence the adsorption of tyrosine as compared to tryptophan. This finding appears somewhat paradox, as hydroxyl groups are known to bind to calcite, which should affect the adsorption position.

## 3.3 Aspartic Acid

Aspartic acid is frequently found in proteins that are important for biomineralisation [29] and contains a second carboxyl group. This is especially interesting, since the carboxyl group has been identified as a suitable group for anchoring the molecule towards calcite surface [12]. Another significant difference is that aspartic acid is considerably smaller than the two amino acids discussed before.

Aspartic acid was sublimated at about 132 °C. Interestingly, upon deposition of around 0.1 ML of aspartic acid onto calcite(104) held at RT, the molecules form

[1]Images with poor resolution did not always display a chiral pattern. However, in case a chiral pattern was seen, it always exhibited the same but mirror-imaged appearance for the D- or L-enantiomers, respectively.

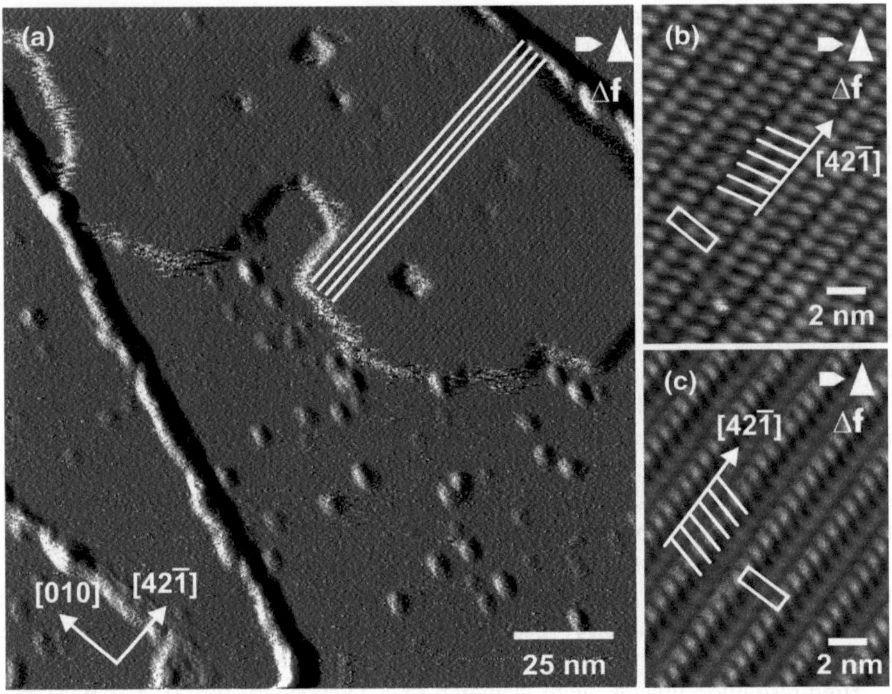


**Fig. 3** Images of **a** D-tyrosine and high-resolution images of **b** D- and **c** L-tyrosine upon deposition onto calcite(104)

islands that appear very similar to the islands formed upon deposition of tyrosine and tryptophan (Fig. 4a). The islands formed by aspartic acid also display an apparent height of about 4 Å and, most surprisingly, a chiral (5 × 1) structure at small scale (Fig. 4b, c). Again, the molecular stripes form an angle of 108° with respect to the [42$\bar{1}$]-direction.

Thus, although the side chain of aspartic acid differs substantially from the side chains of the two molecules considered before, the resulting structure appears to be strikingly similar. This finding might suggest that a general binding motif for amino acids on calcite(104) can be found, which is rather unaffected by the different side chains. This is, however, not true as will be shown by investigating the structures observed for glycine and alanine.

### *3.4 Alanine*

Alanine is the smallest chiral amino acid and, therefore, was used to test whether the above reported (5 × 1) structure constitutes a general binding motif for amino acids on calcite(104). Since its side chain consists of a methyl group, the

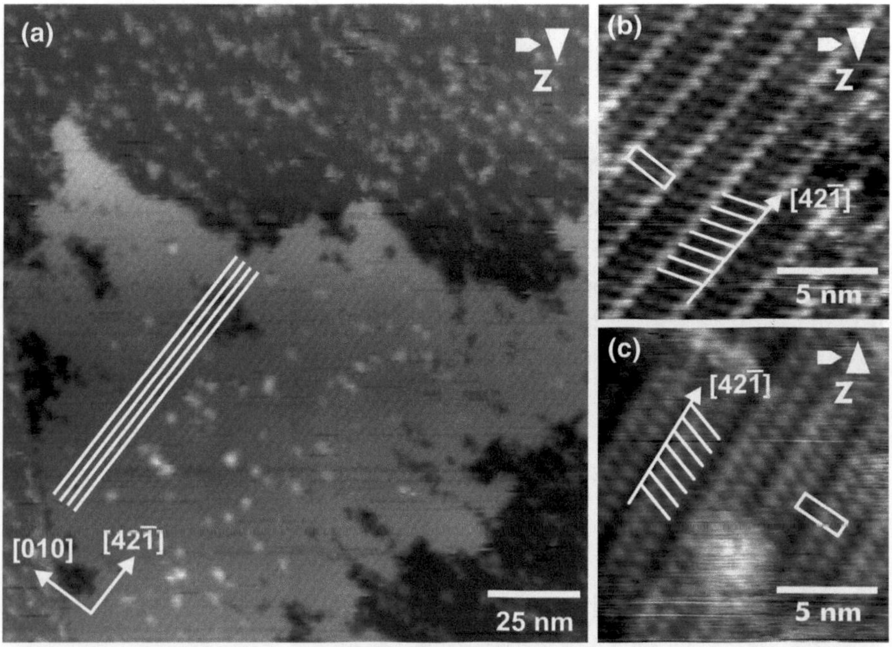


**Fig. 4** Images of **a** L-aspartic acid and high-resolution images of **b** D- and **c** L-aspartic acid upon deposition

intermolecular interactions in the crystal structure are dominated by hydrogen bonds between the amino acid groups [30].

Sublimation at 80 °C for 5–10 min leads to a coverage of about 0.05 ML. Alanine assembles into structures with an apparent height of about 4 Å (Fig. 5). It should be mentioned that the islands are rather small compared to the islands formed from the other amino acids discussed above, where less but larger islands were revealed.

In high-resolution images, a (2 × 2) superstructure can be identified. Probably due to the small size of the structures, no chirality can be observed. As seen for the other amino acids before, the contrasts differ substantially, as demonstrated in the high-resolution images presented in Fig. 5b, c.

## 3.5 Glycine

Glycine, being achiral, consists of the amino acid group only. It, therefore, takes up a special position within the proteinogenic amino acids. Consequently, hydrogen bonds are the dominant intermolecular interactions in the bulk structure of glycine [31].

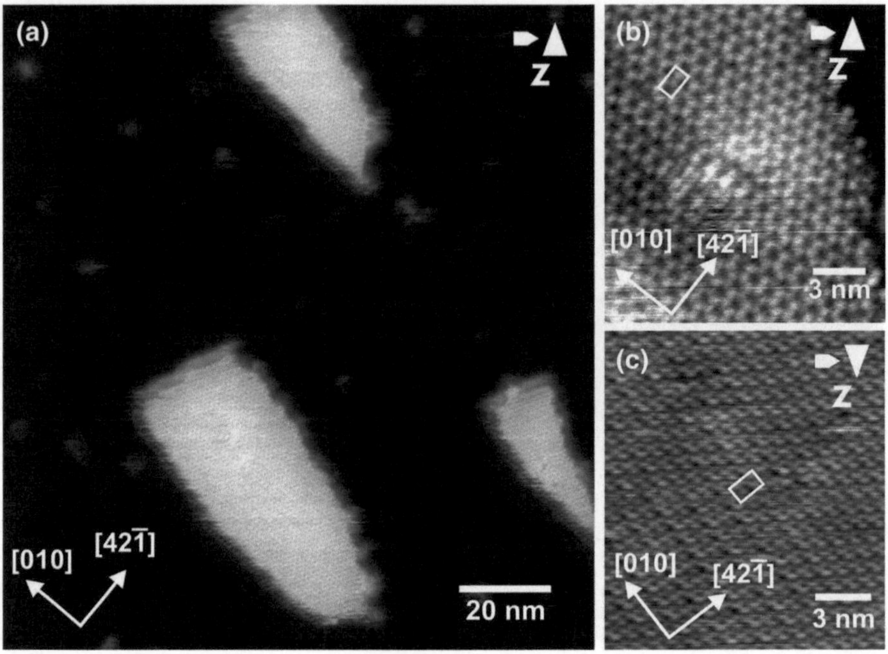


**Fig. 5** **a** D-alanine and high-resolution images of **b** D- and **c** L-alanine upon deposition. The (2 × 2) unit cell is marked with a *rectangle* in **b** and **c**

For sublimation, the cell temperature was about 80 °C, yielding a coverage of about 0.1 ML during 10–30 min of sublimation. Deposition of glycine also results in the formation of islands.

However, compared to the rather well-defined structures that could be observed for the other amino acids, the islands assembled from glycine exhibit a "blurry" appearance with a (1 × 1) superstructure (Fig. 6). Again, the apparent height is about 4 Å. Interestingly, in sharp contrast to what was observed for all other amino acids investigated in this study, the adsorption of glycine results in a metastable structure. This is concluded from the fact that the (1 × 1) islands disappear when waiting overnight, leaving behind unordered structures.

The reason for the transition cannot be identified easily. A possible explanation might be the deprotonation of either the protonated amino or carboxyl group, depending on the state of glycine adsorption on calcite. However, it remains unclear why alanine forms stable islands then, since both molecules are very similar from a chemical point of view.

Most importantly, the smallest amino acid presents yet another adsorption structure, illustrating the richness of binding motifs for amino acids on calcite(104).

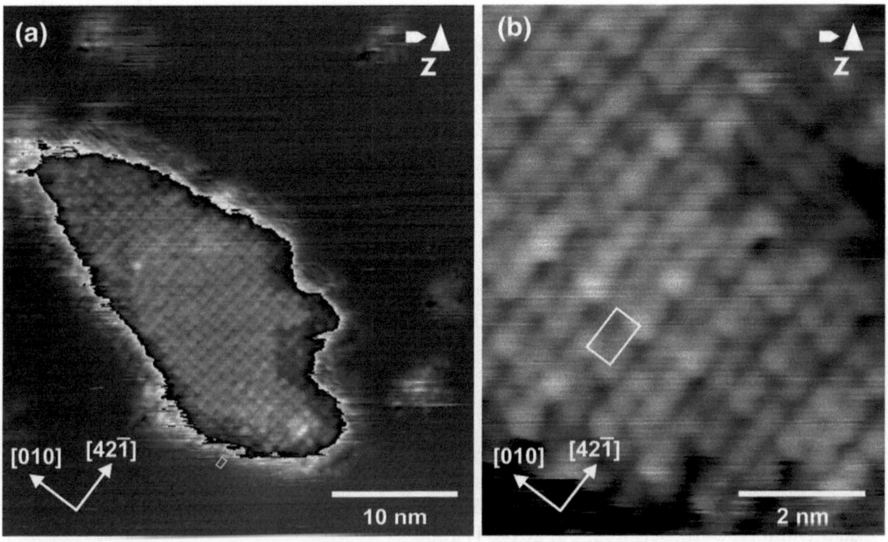


**Fig. 6** Images of glycine deposited onto calcite(104). In **a** a double-colour scale is used to highlight the resolution obtained simultaneously on the calcite substrate and the glycine island. **b** Zoom into the island structure, revealing the (1 × 1) unit cell (marked by a *rectangle*)

# 4 Conclusion

Upon deposition on calcite, tryptophan, tyrosine and aspartic acid self-assemble into island structures. The D- and L-enantiomers form mirrored structures each, as expected from a chiral molecule. Surprisingly, all three molecules exhibit a (5 × 1) superstructure with a very similar inner pattern, despite the significantly different side chains. One possible implication of this finding is that the adsorption geometry is determined mostly by the amino acid group. If this is true, single amino acids are expected to bind in a completely different way to calcite when compared to proteins, since the amino acid moieties are condensed in the latter case. This finding might also suggest that the (5 × 1) superstructure constitutes a general binding motif for amino acids on calcite(104), irrespectively of the chemical nature of the side chains. This conclusion does, however, not hold true as demonstrated by depositing the two smallest amino acids, glycine and alanine, onto calcite(104).

Alanine forms a (2 × 2) superstructure when deposited onto calcite(104) held at RT. Glycine, being even smaller than alanine, assembles into a (1 × 1) superstructure, which appears to constitute a transient structure. These two amino acids demonstrate that a seemingly small change in structure (adding a methyl group when going from glycine to alanine) can cause large changes in the resulting structure. In contrast, the opposite observation is made for the three large amino acids, where large changes in the side chains did not result in different adsorption structures on the surface. Our work, thus, demonstrates that no general binding motifs for the adsorption of amino acids onto the most stable cleavage plane of calcite can be identified, calling for detailed adsorption studies for each amino acid individually.

**Acknowledgments** The authors thank Ralf Bechstein for most stimulating discussions.

## References

1. Miller, S.L.: A production of amino acids under possible primitive earth conditions. Science **117**, 528–529 (1953)
2. Weiner, S., Addadi, L.: Design strategies in mineralized biological materials. J. Mater. Chem. **7**, 689–702 (1997)
3. Schiffrin, A., Reichert, J., Pennec, Y., Auwärter, W., Weber-Bargioni, A., Marschall, M., et al.: Self-assembly of L-methionine on Cu(111): steering chiral organization by substrate reactivity and thermal activation. J. Phys. Chem. C **113**, 12101–12108 (2009)
4. Barlow, S.M., Raval, R.: Complex organic molecules at metal surfaces: bonding, organisation and chirality. Surf. Sci. Rep. **50**, 201–341 (2003)
5. Kühnle, A., Linderoth, T.R., Hammer, B., Besenbacher, F.: Chiral recognition in dimerization of adsorbed crysteine observed by scanning tunneling microscopy. Nature **415**, 891–893 (2002)
6. Barlow, S.M., Louafi, S., Le Roux, D., Williams, J., Muryn, C., Haq, S., Raval, R.: Polymorphism in supramolecular chiral structures of R- and S-alanine on Cu(110). Surf. Sci. **590**, 243–263 (2005)
7. Schiffrin, A., Riemann, A., Auwärter, W., Pennec, Y., Weber-Bargioni, A., Cvetko, D., et al.: Zwitterionic self-assembly of L-methionine nanogratings on the Ag(111) surface. PNAS **104**, 5279–5284 (2007)
8. Forster, M., Dyer, M.S., Persson, M., Raval, R.: Assembly of chiral amino-acids at surfaces from a single molecule perspective: proline on Cu(110). Top. Catal. **54**, 13–19 (2011)
9. Kanazawa, K., Taninaka, A., Huang, H., Nishimura, M., Yoshida, S., Takeuchi, O., et al.: Scanning tunneling microscopy/spectroscopy on self-assembly of a glycine/Cu(111) nanocavity array. Chem. Comm. **47**, 11312–11314 (2011)
10. Orme, C.A., Noy, A., Wierzbicki, A., McBride, M.T., Grantham, M., Teng, H.H., et al.: Formation of chiral morphologies through selective binding of amino acids to calcite surface steps. Nature **411**, 775–779 (2001)
11. Addadi, L., Weiner, S.: Crystals, asymmetry and life. Nature **411**, 753–755 (2001)
12. Rahe, P., Kittelmann, M., Neff, J.L., Nimmrich, M., Reichling, M., Maass, P., Kühnle, A.: Tuning molecular self-assembly on bulk insulator surfaces by anchoring of the organic building blocks. Adv. Mat. **25**, 3948–3956 (2013)
13. Rahe, P., Nimmrich, M., Kühnle, A.: Substrate templating upon self-assembly of hydrogen-bonded molecular networks on an insulating surface. Small **8**, 2969–2977 (2012)
14. Rahe, P., Nimmrich, M., Greuling, A., Schütte, J., Stará, I.G., Rybáček, J., et al.: Toward molecular nanowires self-assembled on an insulating substrate: Heptahelicene-2-carboxylic acid on calcite $(10\bar{1}4)$. J. Phys. Chem. C **114**, 1547–1552 (2010)
15. Reichert, J., Schiffrin, A., Auwärter, W., Weber-Bargioni, A., Marschall, M., Dell'Angela, M., et al.: L-Tyrosine on Ag(111): universality of the amino acid 2D zwitterionic bonding scheme? ACS Nano **4**, 1218–1226 (2010)
16. Wang, D., Xu, Q.-M., Wan, L.-J., Bai, C.-L., Jin, G.: Adsorption of enantiomeric and racemic tyrosine on Cu(111): a scanning tunneling microscopy study. Langmuir **19**, 1958–1962 (2003)

17. Atanasoska, L.L., Buchholz, J.C., Somorjai, G.A.: Low-energy electron diffraction study of the surface structures of adsorbed amino acid monolayers and ordered films deposited on copper crystal surfaces. Surf. Sci. **72**, 189–207 (1978)
18. Zhao, X., Zhao, R.G., Yang, W.S.: Self-assembly of L-tryptophan on the Cu(001) surface. Langmuir **18**, 433–438 (2002)
19. Hao, W., Xue-Ying, Z., Wei-Sheng, Y.: Adsorption of aspartic acid on Cu(001) studied by scanning tunneling microscopy. Acta Phys. Sin. **49**, 1316–1320 (2000)
20. Zheleva, Z.V., Eralp, T., Held, G.: Complete experimental structure determination of the p(3 × 2)pg phase of glycine on Cu{110}. J. Phys. Chem. C **116**, 618–625 (2012)
21. Yitamben, E.N., Niebergall, L., Rankin, R.B., Iski, E.V., Rosenberg, R.A., Greeley, J.P., et al.: Tracking amino acids in chiral quantum corrals. The J. Phys. Chem. C **117**, 11757–11763 (2013)
22. Zhao, X., Wang, H., Zhao, R.G., Yang, W.S.: Self-assembly of amino acids on the Cu(001) surface. Mater. Sci. Eng. C **16**, 41–50 (2001)
23. Rankin, R.B., Sholl, D.S.: Structures of dense glycine and alanine adlayers on chiral Cu (3,1,17) surfaces. Langmuir **22**, 8096–8103 (2006)
24. Clegg, M.L., de la Garza, L.M., Karakatsani, S., King, D.A., Driver, S.M.: Chirality in amino acid overlayers on Cu surfaces. Top. Catal. **54**, 1429–1444 (2011)
25. Rahe, P., Schütte, J., Kühnle, A.: NC-AFM contrast formation on the calcite $(10\bar{1}4)$ surface. J. Phys. Condens. Matter **24**, 084006 (2012)
26. Rahe, P., Bechstein, R., Kühnle, A.: Vertical and lateral drift corrections of scanning probe microscopy images. J. Vac. Sci. Technol. B **28**, C4E31–C4E8 (2010)
27. Hübschle, C.B., Messerschmidt, M., Luger, P.: Crystal structure of DL-Tryptophan at 173K. Cryst. Res. Tech. **39**, 274–278 (2004)
28. Mostad, A., Nissen, H.M., Rømming, C.: Crystal structure of L-tyrosine. Acta Chem. Scand. **26**, 3819–3833 (1972)
29. Weiner, S.: Aspartic acid-rich proteins: major components of the soluble organic matrix of mollusk shells. Calcif. Tissue Int. **29**, 163–167 (1979)
30. Simpson Jr, H.J., Marsh, R.E.: The crystal structure of L-alanine. Acta Crystallogr. **20**, 550–555 (1966)
31. Albrecht, G., Corey, R.B.: The Crystal Structure of Glycine. J. Am. Chem. Soc. **61**, 1087–1103 (1939)

# Origin of the KPFM Contrast in the Adsorption of a Triphenylene Derivative on KBr(001)

**Antoine Hinaut, Florian Chaumeton, David Martrou and Sebastién Gauthier**

**Abstract** The results of the study of the adsorption of 2,3,6,7,10,11-hexacyano-propyloxytriphenylene on KBr(001) by non-contact atomic force microscopy (nc-AFM) coupled to Kelvin probe force microscopy (KPFM) are presented. A particular emphasis is put on the KPFM measurements. In particular, the contrast in the Kelvin potential between a molecular monolayer and the KBr substrate is interpreted on the basis of molecular modeling and classical electrostatic calculations using a spherical tip model. It is demonstrated that this contrast originates from the polarization of the molecule induced by its adsorption conformation where five polar CN groups are adsorbed above five surface $K^+$ ions. The calculated value of the Kelvin potential difference between the molecular monolayer and KBr is in good agreement with the experimental value.

**Keywords** Non-contact atomic force microscopy · Kelvin probe force microscopy · Molecular adsorption on insulating surfaces

## 1 Introduction

Atomic force microscopy in the non-contact (or frequency modulation) mode is now a well-established technique and has proven to be invaluable for investigating molecular adsorption on atomically clean, well-defined surfaces of bulk insulators. This field is progressing rapidly: A wide variety of structures, from 2D or 3D self-assembled islands to single molecules, have been observed on different surfaces with an ever increasing resolution [1–17] and important phenomena have been investigated, such as dewetting of molecular layers from the substrate [3], novel growth modes [15], and on-surface synthesis of polymeric structures by thermally induced covalent linking [13].

A. Hinaut · F. Chaumeton · D. Martrou · S. Gauthier (✉)
CNRS, CEMES (Centre d'Elaboration des Matériaux et d'Etudes Structurales),
BP 94347, 29 Rue Jeanne Marvig, 31055 Toulouse, France
e-mail: gauthier@cemes.fr



P. Moriarty and S. Gauthier (eds.), *Imaging and Manipulation of Adsorbates Using Dynamic Force Microscopy*, Advances in Atom and Single Molecule Machines,
DOI 10.1007/978-3-319-17401-3_9

During the same period, Kelvin probe force microscopy (KPFM) has been combined with nc-AFM [18–22] to investigate metallic or semiconducting surfaces, as well as adsorbates [23, 24] or thin insulating films on metals [20, 25, 26]. But the application of this technique to bulk insulating surfaces [27–31] is only beginning. Coupling these two techniques is not only interesting to characterize the electrical properties of the adsorbates, but also to extract topographic images that are free from electrostatic-forces-induced distortion [37, 38]. A key difficulty with KPFM on insulators, however, is that the interpretation of the results is not yet well established and for this reason remains only qualitative.

In the following, our goal was to provide a semi-quantitative interpretation of the KPFM results recently obtained for the adsorption of hexacyanopropyloxytriphenylene (HCPTP) on KBr(001) [30]. This molecule (Fig. 1) was equipped with six flexible propyl chains ending with dipolar CN groups in order to increase its adsorption energy and to limit its diffusion at room temperature. It was proposed [30] that in one of the two molecular monolayer structures identified on KBr, the molecules are lying flat on the surface, with five of these CN groups interacting with surface $K^+$ ions. This adsorption conformation polarizes the molecule, and the primary question we address in this chapter is how the Kelvin voltage measured above these monolayers can be related to this static polarization. While the quantitative aspects of KPFM on metals have been the focus of a number of recent works [20, 32, 33], no real attempts at a quantitative analysis have been reported yet for molecular deposits on bulk insulators. Note that we are not here concerned with the recent reports and analysis of atomic resolution KPFM on insulators [34, 35] where the data are obtained in a different regime of much stronger tip–substrate interaction, where short-range electrostatic forces play a dominant role.

In the following, the experimental results obtained on HCPTP/KBr(001) are briefly recalled in Sect. 2, with an emphasis on the KPFM measurements. The

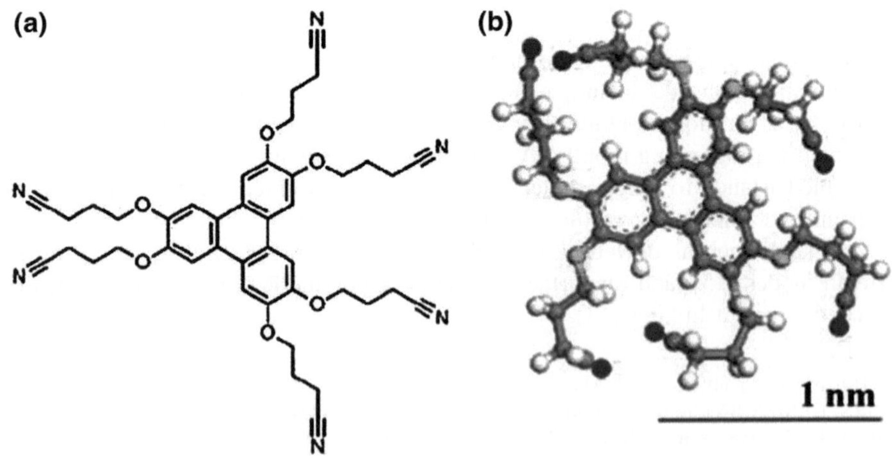


**Fig. 1** **a** Molecular scheme and **b** structure of HCPTP optimized in vacuum

interpretation of these measurements is then presented in Sect. 3, based on the calculation method of the Kelvin potential that is described in Annex A. These results are discussed in Sect. 4. In addition, an estimation of the tip-surface force that is used to fix some parameters of the calculations is presented in Annex B.

## 2 Experimental Results

### *2.1 Experimental Details*

Experiments were conducted with a commercial room temperature ultra-high vacuum STM/AFM (Oxford Instruments Omicron NanoScience, Taunusstein, Germany).

The KBr crystal was cleaved in air, transferred to the UHV system, and heated at 480 K for 1 h to remove the charges created during the cleavage process. This preparation method results in an atomically well-ordered surface with (001) terraces separated by atomic steps, mainly oriented in the nonpolar ⟨100⟩ crystallographic directions. Molecules were deposited from a heated boron nitride disk on which a few drops of the molecular solution have been let dry. QNCHR (uncoated) and PtNCH (Pt-coated) silicon cantilevers from NanoSensors (Neuchatel, Switzerland) were used. No special preparation except a moderate heating (420 K) in vacuum was exercised. The resonance frequencies were close to 300 kHz, with quality factors around 40,000.

KPFM was performed in the frequency modulation mode [19]. The bias voltage, referred to the sample, was modulated at 1 kHz with an amplitude of 2 V.

The images were obtained in the constant $\Delta f$ mode with an oscillation amplitude $A = 2$ nm and small values of $\Delta f$, corresponding to normalized frequency shifts [36] $\gamma = kA^{3/2}\Delta f/f_0 = 0.25$ fN m$^{1/2}$ at most ($k \sim 40$ N m$^{-1}$). In these conditions, the interaction of the tip with the surface is quite weak. It is then expected that only van der Waals and electrostatic forces contribute to the image (see Annex 2). In these conditions and at the Kelvin voltage, the topographic image is close to a pure van der Waals image [37, 38]. The apparent height of the observed structures is then close to the values extracted from the molecular models, facilitating the structural identification.

Great caution was exercised to avoid cross talk between the different signal channels available in nc-AFM. Constant $\Delta f$ images were recorded simultaneously with maps of the Kelvin voltage, the frequency shift, the amplitude, and the excitation voltage. The excitation voltage map was uniform, at a value close to its value in the absence of tip–surface interaction for all the images of this paper. The horizontal and vertical directions in the presented images are aligned on ⟨100⟩ nonpolar directions of KBr.

## 2.2 Results

A topographic and a Kelvin image of a KBr sample after deposition of molecules at room temperature and annealing at 420 K during 30 min is presented in Fig. 2. On the basis of our previous work [30], we can identify three different domains: (1) the two horizontally elongated areas that appear near the center of image (a) reveal the KBr surface. They are decorated by small molecular aggregates which are probably anchored on defects; (2) other areas, labeled MLh, which present fewer defects are associated with a monomolecular layer where the molecules lie flat on the KBr surface; (3) finally, the areas labeled MLv, which exhibit stripes oriented in polar <110> directions of the substrate correspond to molecular structures where the molecules are standing approximately perpendicular to the surface. These attributions are consistent with the measured height of the two types of molecular layers: The apparent height of MLh is 0.4 nm, a reasonable value for a molecule lying flat on the surface, while that of MLv is 1.6 nm, a value comparable to the molecular diameter (see Fig. 1).

These structures form, during a dewetting process where molecules are transferred from MLh to MLv, liberating KBr patches, since molecular density is higher in MLv than in MLh. This type of dewetting process is common in systems where the molecules are weakly bound to the surface and has already been observed during nc-AFM experiments [3, 39]. We suppose that the anchored molecular aggregates that are observed on KBr are formed during this process, at the annealing temperature of 420 K, but a better understanding of this phenomenon requires further experiments.

The Kelvin map of Fig. 2b shows a very clear contrast between KBr and the two types of molecular layers. As shown by the upper profile of Fig. 3, the Kelvin

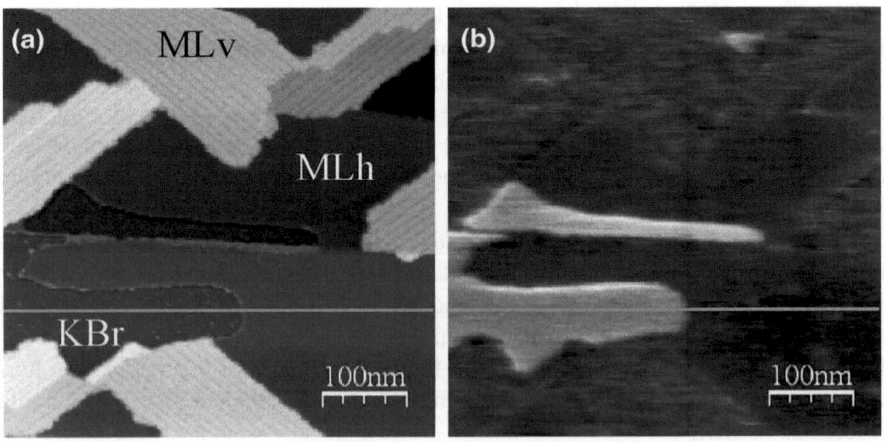


**Fig. 2** Images of a sample annealed at 420 K after the deposition of the molecules at room temperature. **a** Topographic image and **b** simultaneously obtained Kelvin map. Imaging conditions: $A = 2$ nm, $\Delta f = -20$ Hz, $f_{mod} = 1$ kHz, $U_{mod} = 2$ V

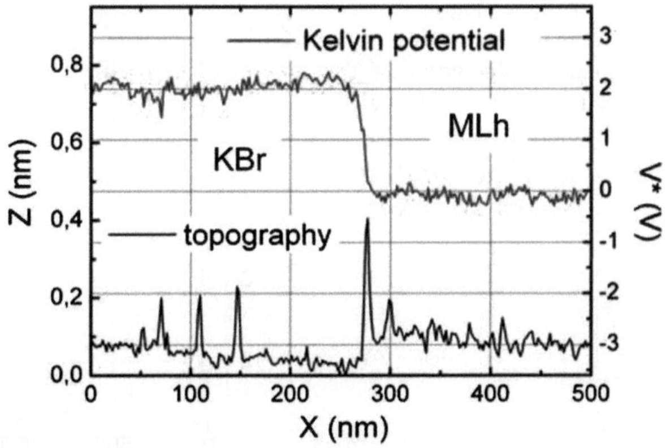


**Fig. 3** Profiles corresponding to the line drawn in Fig. 2a (*black profile*), b (*red profile*). The spikes that appear on the topographic profile on KBr are related to the molecular aggregates mentioned in the text (color figure online)

voltage is shifted down by about 2 V from KBr to MLh. The average Kelvin voltage on MLv is higher than on MLh. It is also more heterogeneous on MLv, revealing a disorder that is not so visible in the topographic image. In the following, we will concentrate on the interpretation of the Kelvin voltage on KBr and on MLh only.

Note that the two KBr patches that appear in Fig. 2 are bordered by a KBr monoatomic step (height 0.33 nm), reducing the apparent height difference between KBr and the adjacent MLh layer to approximately 0.1 nm, as measured on the topographic profile of Fig. 3.

Measurements of the Kelvin potential on the sample imaged in Fig. 2, during more than one day and a half, are displayed in Fig. 4. The black squares give the value on the MLh layer. The sequence of voltage steps that appear near 1100 min is related to a sequence of images where higher spatial resolution was looked for and consequently where harsher imaging conditions were used ($\Delta f$ down to $-80$ Hz, KPFM feedback loop opened). During this sequence, the tip evolved, as seen from the topographic images. Accordingly, tip changes are clearly the origin of the Kelvin voltage steps, as observed and analyzed in Ref. [40]. Before and after these evolutions, the surface was imaged with the KPFM loop active in mild conditions ($\Delta f = -20$ Hz). It is then observed that the absolute value of the Kelvin voltage decreases exponentially, with a time constant of 360 min (blue lines in Fig. 4).

A similar phenomenon is commonly observed on the clean surfaces of ionic crystals (alkali halides, MgO, etc.). It is related to the relaxation of the charges created by the cleavage of the crystal during the preparation of the surface [27, 28]. Here, the situation is different: The molecules were deposited at room temperature on the surface of a crystal that was previously annealed to 480 K in order to remove these surface charges, resulting in Kelvin voltages close to $0 \pm 1$ V. It is only after annealing the sample covered by molecules that a large negative Kelvin voltage was observed. This observation points to a large charge rearrangement during this step, but we lack an understanding of this phenomenon at the present time.

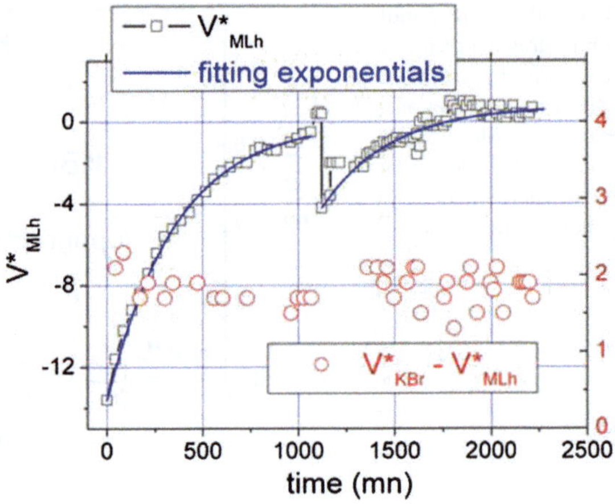


**Fig. 4** Time evolution of the Kelvin potential on MLh ($V^*_{\mathrm{MLh}}$) and of the difference between the Kelvin potential on KBr and on MLh ($V^*_{\mathrm{KBr}} - V^*_{\mathrm{MLh}}$)

An important point illustrated by Fig. 4 is the absence of correlation between the absolute value of the observed Kelvin voltage and the Kelvin voltage difference between MLh and KBr. Clearly, the evolutions of these two curves are associated with two apparently disconnected phenomena. Indeed, it was demonstrated in Ref. [40] that the mean value of the Kelvin voltage is largely fixed by the electrostatic characteristics of the tip. This is not the case for the relative values of the Kevin voltage, which are directly related to the electrostatic properties of the surface phases under examination.

In the following, we concentrate on the interpretation of the value of the Kelvin voltage difference $\Delta V^*$ between MLh and KBr. This value was measured on different samples with 3 different tips, 2 uncoated Si tips (QNCHR), and one Pt-coated tip (PtNCH). Care was taken to minimize the influence of the edges of the molecular islands by evaluating the Kelvin potential at a minimal distance of approximately 50 nm from them. We obtain $\Delta V^* = 1.8$ V ± 0.2 V, with no clear systematic difference between the tips.[1]

## 2.3 MLh Structure

Materials Studio [41] was used with the COMPASS force field [42] to optimize the conformation of an isolated molecule on KBr(001). The condensed-phase optimized molecular potentials for atomistic simulation studies (COMPASS) force field is well adapted to calculate the adsorption of organic molecules on the surface of inorganic materials because it has been parameterized using condensed-phase

[1]The value of 1.4 ± 0.2 V given in our previous work (Ref. [30]) was the difference between the Kelvin voltage on KBr and an average of the voltage on MLh and MLv. As the Kelvin voltage is higher on MLv, this explains the difference with the value we obtain here.

properties as well as ab initio and empirical data for isolated molecules [43, 44]. Previous studies have shown that the interaction of organic molecules with ionic surfaces is dominated by van der Waals and electrostatic forces. In addition, these works indicate that the charge transfer between the substrate and the molecule is negligible [45], ensuring that force-field-based calculations are sufficient for a realistic description of these systems. The KBr slab was constituted of 6 × 6 × 3 unit cells, corresponding to a thickness of six (001) atomic layers. The two upper layers were free to relax during the simulations while the four lower layers were kept fixed.

The calculations show that the molecule binds on KBr(001) by positioning five of its CN groups above $K^+$ ions, with a mean N–$K^+$ distance of 0.28 nm. The last CN group cannot bind to the surface and stays at a larger distance from the surface (see Fig. 5 and the arrow in Fig. 6). In this conformation, the central aromatic core of the molecule is lying flat at 0.4 nm from the surface plane.

The adsorption energy amounts to 1.8 eV, including not only the contribution of the five CN groups but also the interaction energy of the oxygen atoms and the aromatic core with the surface. This contribution can be roughly evaluated from the calculated adsorption energy of hexamethoxytriphenylene on KBr(001) in the same conditions, which amounts to 0.8 eV. We conclude that each CN group contributes approximately to (1.8 − 0.8)/5 = 0.2 eV, in fair agreement with the value obtained for the CN groups of the truxene derivative calculated in [9].

In our previous work [30], we proposed a tentative model for the structure of the MLh layer, taking into account the parameters of the molecular lattice, measured from high-resolution images. As we will show in the following, it was impossible with this model to find reasonable agreement with the calculated values of the Kelvin voltage. The model was based on two assumptions: (1) The molecules in the monolayer are adsorbed in the same configuration as the isolated molecule and (2) there are two molecules in each unit cell. While (1) seems reasonable, since these molecules are not expected to interact strongly when lying flat on a surface,

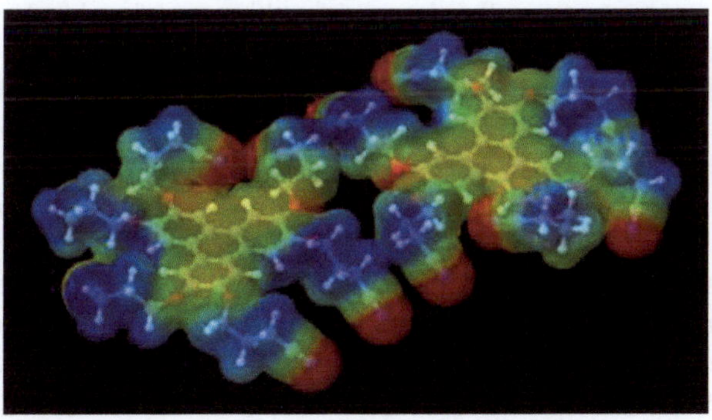

**Fig. 5** Two HCPTP molecules in their adsorption conformation. The molecular van der Waals surface is colored according to the molecular electrostatic potential from *blue* (more positive) to *red* (more negative)

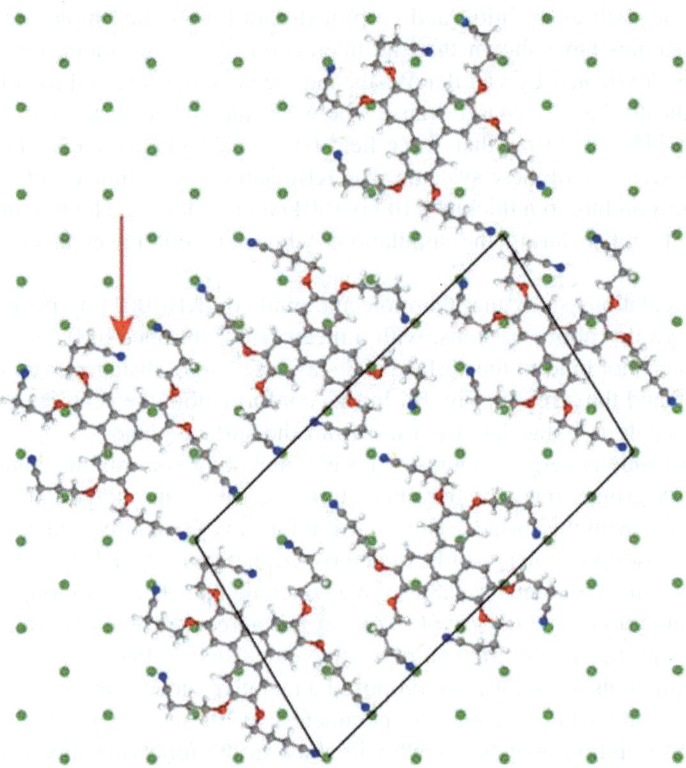

**Fig. 6** Model of the MLh layer used to evaluate the Kelvin potential. The *green dots* represent $K^+$ ions of the KBr(001) surface. For each molecule, 5 CN groups are electrostatically bound to a surface $K^+$ ion, while the sixth CN group (*red arrow*) does not interact strongly with the surface. The unit cell area is 6.08 $nm^2$

(2) is probably wrong since it leads to a molecular density that is much lower than the compact arrangement usually observed for this type of molecular layer. But, we did not succeed in finding a denser model respecting (1) that would be compatible with the experimentally determined lattice parameters. In the real structure, the molecules are probably slightly distorted from their isolated molecule configuration to achieve a higher density. This structure could be in principle obtained by calculations, but the energy differences between the different configurations involved are too small for the COMPASS force field to remain reliable.

Once (1) is accepted, the only parameter that influences the value of the Kelvin voltage and that differs from one structure to the other is the molecular density. The MLh structure proposed in Fig. 6 combines molecules in their isolated configuration while achieving a molecular density 1.64 higher than the previously proposed model. It does not correspond to the real structure since its lattice parameters are different from the experimentally determined ones. Nevertheless, we think that it constitutes a reasonable starting point to evaluate realistically the Kelvin potential.

In addition, we foresee that even denser molecular arrangements could be achieved since some of the molecules in this model are still far from being in van der Waals contact.

## 3 Estimation of the Kelvin Voltage Difference Between KBr and the Molecular Monolayer MLh

The major difficulty in the interpretation of KPFM, and, more generally, of local probe electrostatic imaging methods, is related to the long-range character of the electrostatic forces. (This discussion does not apply to atomic scale KPFM where the forces involved, e.g., polarization forces, are of much shorter range [34, 35]). This long-range interaction makes the images dependent on the shape of the tip [46–49], which is generally unknown. Most of these experiments are carried in ambient conditions and use large oscillations amplitudes and large tip–substrate distances. In these conditions, not only the trunk of the tip but even the supporting cantilever can also contribute to the image [48, 49]. The situation is somewhat different for many UHV experiments where both smaller amplitudes and smaller tip–substrate distances can be used. This is even more the case for FM-KPFM, which, using the electrostatic force gradient, has a better resolution or in other words a more local character. It is thus expected that the KPFM images obtained under these conditions will be dominated by the near apex region of the tip. In the following, we use a simple spherical tip–plane substrate model, which has the advantage of leading easily to approximate analytical expressions [40, 50] and is simple to numerically implement by means of the image charge method.

Our calculations are based purely on a classical electrostatic model that is presented in Annex A. The atomistic nature of the tip and the substrate is ignored. Molecules are introduced in the tip–substrate junction as distributions of partial atomic charges corresponding to the minimal energy configuration of a single molecule adsorbed on KBr(001) calculated previously. KBr is characterized by its relative static dielectric constant $\varepsilon = 4.87$.

The Kelvin potential $U_K^{AM}$ (respectively, $U_K^{FM}$) is calculated for the two commonly used KPFM modes, the AM-KPFM (respectively, FM-KPFM) mode where the Kelvin feedback loop minimizes the electrostatic force (respectively, the gradient of the electrostatic force).

The calculated constant height image[2] of the Kelvin potential $U_K^{FM}$ over a molecular island of HCPTP on KBr made by a $20 \times 20$ repetition of the unit cell displayed in Fig. 6 is shown in Fig. 7. Calculated constant height profiles of the Kelvin potential obtained on a central line, as in Fig. 7 and as a function of the

[2]Due to the weak distance dependence of $U_k$, constant $\Delta f$ Kelvin images (such as the experimental image in Fig. 2b) do not differ significantly from constant height images (such as the calculated one in Fig. 7).

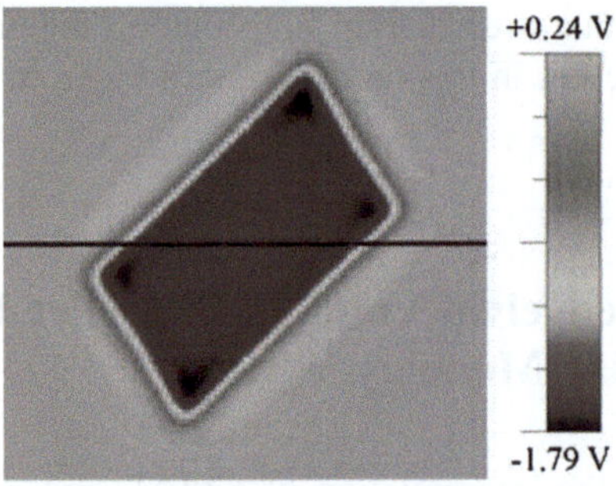


**Fig. 7** Calculated Kelvin potential $U_K^{FM}$ over a 20 × 20 unit cells island (800 molecules) on KBr. The black line corresponds to the $n$ = 20 profile in Fig. 8c

island size, are displayed in Fig. 8. The radius $R$ = 5 nm and the closest distance between the surface of the sphere and the KBr surface $b$ = 2 nm were chosen to fit approximately with the parameters extracted from the force curves analyzed in Annex B.

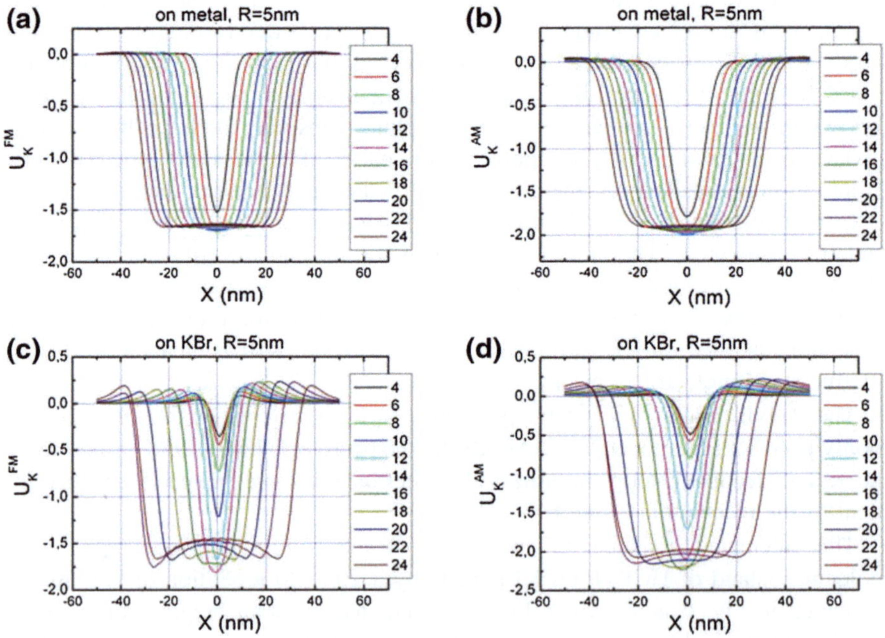


**Fig. 8** Constant height profiles of the Kelvin potential on an island of HCPTP molecules as a function of its $n \times n$ size; **a** and **b** on a metal substrate, **c** and **d** on KBr. $R$ = 5 nm, $b$ = 2 nm. **a** and **c** (Resp. **b** and **d**) give $U_K^{FM}$ (Resp. $U_K^{AM}$), the FM (respectively, AM)-KPFM Kelvin potential

Not surprisingly, the calculated behavior is similar to what is observed on a dipolar layer in Annex A. On the metal substrate, the FM-KPFM Kelvin potential $U_K^{FM}$ near the center of the island (Fig. 8a) is nearly constant around −1.65 V, as soon as the size of the island exceeds the size of the tip. For the AM-KPFM Kelvin potential $U_K^{AM}$ (Fig. 8b), the convergence is much slower, the value reaching −1.89 V for the largest island, whose size is around 60 nm. This confirms the lower spatial resolution of AM-KPFM: the tip "feels" the edges of even the largest island.

The z component of the dipole moment of the adsorbed molecule, extracted from the previous calculation, is $p = -0.2734 \text{ nm} \times q = -13.1$ Debye and the surface of the unit cell, containing two molecules, is $A = 6.08 \text{ nm}^2$ (Fig. 6). The potential drop associated with the dipolar layer of dipolar density $p/A$ and its image in the substrate is given by: $\Delta V_m = p/(\varepsilon_0 \cdot A/2) = -1.625$ V. This is the value that would be measured on a molecular monolayer with the tip replaced by a planar electrode or with the usual "macroscopic" Kelvin probe method, based on a vibrating capacitor [51, 52]. As expected, it is quite close to the value calculated for $U_K^{FM}$ above the center of large enough molecular islands, a situation in which the tip–substrate junction can be considered as a planar junction.

On KBr, the same trends are observable. Near the center of the islands $U_K^{FM}$ converges toward $\Delta V_d = -1.46$ V while $U_K^{AM}$ reaches −1.98 V. $\Delta V_d$ is close to the value predicted by expression 4 of Annex A: $\Delta V_d = [(1 + \beta)/2]\, \Delta V_m = -1.35$ V, with $\beta = (\varepsilon - 1)/(\varepsilon + 2)$, confirming that this reduced value is due to the less efficient electrostatic screening of KBr.

As discussed in Annex A, the precise shape of the profiles is dictated by the spherical shape of the tip and for this reason should not be considered as realistic.

## 4 Discussion and Conclusion

The value given by the calculation is finally $\Delta V_d = -1.46$ V for the structure illustrated in Fig. 6. This value is lower than the experimental value $\Delta V^* = -1.8 \pm 0.2$ V, but we nevertheless consider this result as satisfactory since the structure of the monolayer is not precisely known. It is clear from Fig. 6 that it could be denser and that this would improve the agreement provided which the molecule keeps staying with five CN legs on $K^+$ ions. Indeed, it is possible to find denser structures with moderate van der Waals overlaps between neighboring molecules that give $\Delta V = -1.64$ V. The important point is not to get precise values but to identify the parameters that influence the value of the measured KPFM voltage. This comparison with the experiment shows that the major quantity is the dipolar density of the molecular layer. It also shows that the screening by the substrate can be taken into account by generalizing the usual relation $\Delta V_m = p/(\varepsilon_0 \cdot A)$ which is valid on a metal to $\Delta V_d = (1 + \beta)\, p/(2 \cdot \varepsilon_0 \cdot A)$.

The calculations presented in Annex A show than even in a very favorable situation (FM-KPFM, small amplitude and small tip–surface distance), modeling the tip as a sphere could be a limiting approximation. Calculating the Kelvin potential with more elaborate tip shapes does not constitute a problem. Many methods have been devised for that purpose. The problem is to find experimental procedures to determine the shape of the tip. Such procedures have been proposed, based for instance on $\Delta f(z)$ spectra in the van der Waals regime (as done here in Annex B) or in the presence of the electrostatic force [53]. As shown in Annex A, the shape of the Kelvin profiles in the vicinity of island edges is determined by the shape of the tip. This is especially marked on the insulating substrate. It could then be possible to use the experimental profiles to improve our knowledge of the tip.

**Acknowledgments** This work has been supported by the European Commission within the projects ARTIST (Contract n° 243421) and AtMol (Contract n° ICT-270028).

## Annex A: Calculation of the Kelvin Voltage

### *Introduction*

We consider a spherical metallic tip in front of a thick insulating substrate fixed on a metallic sample plate. "Thick" means here that the substrate can be considered as semi-infinite. We will investigate the influence on the Kelvin voltage of the introduction of charges into this junction. These charges can be linked to the substrate (e.g., to model a molecule by its partial charges) or to the tip (e.g., to model an insulating cluster adsorbed on the tip: a nano-tip). The case of a metallic substrate, treated by several authors [40, 54–56], will be also considered for comparison. The calculation is classical and does not take into account the atomistic nature of the tip and the substrate, which are considered as continuous solids, characterized by their static dielectric constant. The dipolar layer that is present at the surface of the metallic tip (with an extension in the 0.1 nm range) and the Madelung potential that is present at the surface of the ionic crystal will be ignored. Consequently, the conclusions of these calculations should be considered with great caution when the charges are located very close to the tip and the substrate. Another limitation is that no structural relaxations are taken into account in the calculation. These relaxations have been shown to play an essential role in the formation of atomically resolved KPFM images [34], but they are expected to be negligible in our case where the tip interacts only by weak electrostatic and van der Waals forces with the surface (see Annex B for an evaluation of these forces).

## Derivation of the Kelvin Potential

The energy of the junction formed by the back plate supporting the insulating sample and the tip can be derived in the presence of a charge $q$ by applying the superposition principle (following the notation of Ref. [40]):

$$E = \frac{1}{2} q\Phi_q(r, r_t) + q\Phi_C(r, r_t, U) - \frac{1}{2} C(z_t)U^2 \quad (1)$$

or

$$E = u_0 + u_1 U + u_2 U^2$$

$U$ is the applied bias voltage, $r = (x, y, z)$ labels the position of the charge, and $r_t = (x_t, y_t, z_t)$ the position of the tip. This expression was derived in Ref. [54] for a metallic substrate, but it remains valid in the presence of an insulating substrate. When the charge is introduced into the junction, it polarizes the tip and the sample. This polarization generates the potential $\Phi_q$ which interacts with the charge itself (note the factor of ½). This first term does not depend on $U$. The third term is the energy of the capacitor, including the work delivered by the power supply to charge it (note the negative sign). The second term describes the interaction of the charge with the potential generated by the capacitor biased at $U$, which is proportional to $U$. Note that the first term varies with $q^2$, while the last one does not depend on $q$. The sign of the charge, which is the first information one wants to get, is determined by the second term.

From expression (1), the electrostatic force on the tip reads:

$$F = -\frac{\partial E}{\partial z} = -u_0' - u_1' U - u_2' U^2$$

The AM-KPFM feedback loop minimizes *the force* with respect to the bias $U$ [19]:

$$\frac{\partial F_z}{\partial U} = 0 \Rightarrow u_1' + 2u_2' U_K^{AM} = 0 \Rightarrow U_K^{AM} = -\frac{u_1'}{2u_2'} \quad (2)$$

The FM-KPFM feedback loop minimizes the *force gradient* with respect to the bias U [19]:

$$U_K^{FM} = -\frac{u_1''}{2u_2''} \quad (3)$$

These expressions also appear, in a different form, in the work of Borowik et al. [57]. They show that it is sufficient to evaluate $u_1$, that is, the capacitive potential at the position of the charge and $u_2$, the tip–substrate capacitance, to calculate the Kelvin potential $U_K$.

The generalization to several charges is straightforward. The mutual interaction between the charges (direct or mediated by the electrodes) adds new terms to the energy [Eq. (1)], but these terms do not affect the value of $U_K$, since provided that all the charges are linked to the sample or the tip, they do not depend on $z$. Expressions (2) and (3) remain valid, with: $u_1 = \sum_i q_i \phi_C(r_i, r_t, U)$. $U_K$ is then additive: $U_K = \sum_i U_K^i$, where $U_K^i$ is the Kelvin potential calculated with only the charge i in the tip–sample junction.

Expression (2) and (3) are valid only in the limit of small oscillation amplitude ($U_K$ as given by (2) or (3) should not vary too much over an oscillation cycle). They can be simply generalized by averaging them over an oscillation cycle, as reported in Ref. [54].

Note that the contact potential difference $U_{CPD}$ that exists between the sample plate and the tip has been neglected in this presentation since its effect is only to shift $U_K$ to $U_K - U_{CPD}$.

## *Calculation by the Method of Images*

An approximate analytical expression exists for the metallic sphere-metallic substrate capacitance [58], but not for the metallic sphere–dielectric substrate capacitance. Since the analytical expressions for image charge in a metallic sphere and in a semi-infinite planar dielectric are known [59, 60], it is possible to express the metallic sphere-dielectric plane capacitance as an infinite sum of image charges [46, 50, 60, 61]. When charges are introduced into the tip–substrate junction, it is straightforward to calculate the series of image charges in both electrodes. It turns out that as demonstrated in [50, 54], the resulting series converge relatively rapidly making the calculation easily tractable, even with a large number (e.g., thousands) of charges.

The KPFM profiles $U_K^{FM}$ and $U_K^{AM}$ shown in Fig. 9 have been obtained with a tip of radius $R = 5$ nm, scanning at a tip–surface distance $b = 1$ nm relative to the KBr substrate above square islands constituted by the periodic repetition of a doublet of charges on a square lattice of parameter 1 nm. The doublet of charge ($z_1 = 0.1$ nm, $q_1 = q$) ($z_2 = 0.11$ nm, $q_2 = -q$) is supposed to approximate a dipole ($q$ is the positive elementary charge). The substrate is a metal for Fig. 9a, b and KBr for Fig. 9c, d. $L$ is the length of a side of the square island.

It is immediately seen that the Kelvin potential is positive over the molecular island. This observation confirms the already established rule [62]: A negative charge, or a dipole whose negative end is toward the tip, shifts the Kelvin voltage toward positive values. The value of the Kelvin potential does not very considerably from one case to the other: It lies between 0.15 and 0.2 V. Another general observation is that the observed features are sharper in the FM mode than in the AM

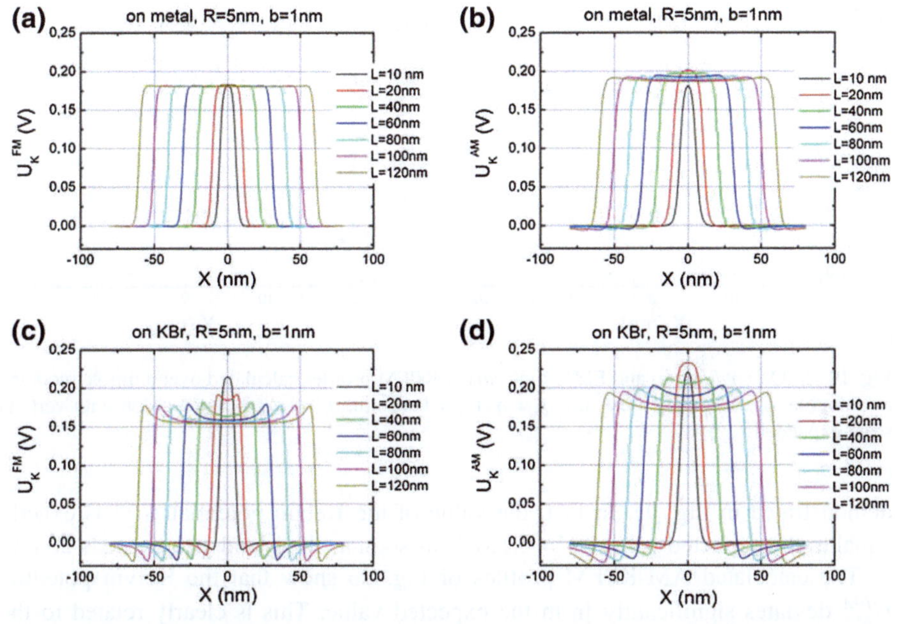


**Fig. 9** FM (**a** and **c**)- and AM (**b** and **d**)-KPFM profiles calculated over a dipolar island on a metal (**a** and **c**) and a KBr (**c** and **d**) surface, as described in the text. Each profile color corresponds to a different island size as indicated in the legend

mode, a direct consequence of monitoring the force gradient in FM mode instead of the force in AM mode.

One surprising feature is the shape of the profiles above the edges of the islands, especially on the KBr substrate. It turns out that the Kelvin potential of a single dipole changes its sign when the lateral distance between this object and the tip becomes larger than approximately a tip radius. This is illustrated in Fig. 10. This particular shape does not result from an imprecise calculation. It is also observed when the calculation is performed with other types of tips, for instance a sphere connected to a conical trunk [63]. We have no simple explanation for this effect, which is very weak but significant when summed over all the dipoles of an island, as seen in Fig. 9.

These calculations show that the only situation in which the Kelvin potential can be realistically calculated with the spherical tip model is when the tip is close to the center of a sufficiently large island. In these conditions, the values of $U_K$ converge toward a constant value.

If the tip was replaced by a planar electrode, the Kelvin potential would be given by the potential drop across the dipole layer: $\Delta V_m = p/(\varepsilon_0 \cdot A) = 0.181$ V, with our values $p = 0.01$ nm $\times\, q$ and $A = 1$ nm$^2$. One expects to recover a similar value when the junction can be approximated as a planar capacitor. This is indeed the case as

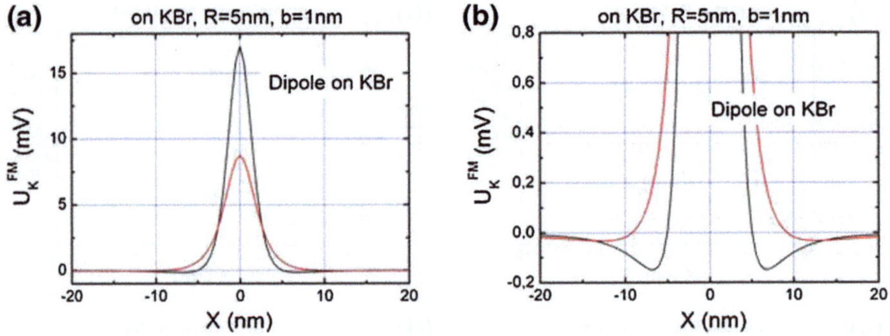


**Fig. 10** **a** AM (*red curve*) and FM (*black curve*)-KPFM profiles calculated over a dipole modeled as a doublet of charge ($z_1$ = 0.1 nm, $q_1 = q$) ($z_2$ = 0.11 nm, $q_2 = -q$). **b** Same graph with vertical scale expanded

demonstrated in Fig. 9a. In fact, the value of the Kelvin potential $U_K^{FM}$ is exactly equal to the expected value $\Delta V = 0.181$ V as soon as the island size exceeds 20 nm.

The calculated AM-KPFM profiles of Fig. 9b show that the Kelvin potential $U_K^{AM}$ deviates significantly from the expected value. This is clearly related to the less local character of the AM mode, as discussed in [19, 20], which is also responsible for the lower spatial resolution of this technique. Nevertheless, in the center of the island, $U_K^{AM}$ still converges toward the theoretical value, being equal to 0.188 V for the largest island calculated.

On KBr (Fig. 9c) $U_K^{FM}$ above the center of the island converges toward 0.154 V. This lower value is related to the less efficient screening of the dielectric substrate when compared to the metal: While a dipole $p$ on the metal develops an image dipole $p$ in the metal giving a total dipole of $P_m = 2p$, the image dipole in the dielectrics is only $\beta p$, with $\beta = (\varepsilon - 1)/(\varepsilon + 1)$, giving $P_d = (1 + \beta)p$. One then expects to get:

$$\Delta V_d = P_d/P_m \quad \Delta V_m = [(1+\beta)/2] \quad \Delta V_m = 0.150 \text{ V} \tag{4}$$

which is quite close to the calculated value. $U_K^{AM}$ on KBr also converges toward this value, but as observed on the metal substrate in a much slower way.

## Annex B: Estimation of the van der Waals Tip-Sample Force

$\Delta f(z)$ spectra were recorded on the sample of Fig. 2 with a QNCHR cantilever. They were obtained with an applied bias corresponding to the Kelvin potential. It is then expected that the tip–surface interaction will be dominated by van der Waals forces if the tip is sufficiently far from the surface for the short-range chemical forces to be negligible.

The $\Delta f(z)$ spectra of Fig. 12 were recorded on a KBr and MLh regions, as indicated in Fig. 11. They can be very well fitted with the $\Delta f$ expression for a sphere-plane van der Waals force [64]:

$$\Delta f = -HRf_0/\left\{6k\left[(d_0 + A_0)^2 - A_0^2\right]^{3/2}\right\} \tag{5}$$

where $H$ is the Hamaker constant of the tip–vacuum–substrate junction. Two parameters are involved in the adjustment. $P_1 = HR\, f_0/6k$ which determines the $\Delta f$ scale and $P_2$, which fixes the origin of the tip–substrate distance scale, which is not accessible experimentally. The relation between the experimental and physical scale is given by: $z + P_2 = d_0 + A_0$.

The fit of the $\Delta f(z)$ spectrum of Fig. 12a, obtained on KBr, is excellent ($R^2 = 0.999$). Taking $k = 42\ \mathrm{N\ m^{-1}}$, the value given by Nanosensors, one obtains:

$$HR = -6kP_1/f_0 = 2.6 \times 10^{-28}\,\mathrm{N\,m^2}$$

It is likely that the tip is covered with KBr. Then, $H = 5.6 \times 10^{-20}$ J, the value of the Hamaker constant for KBr calculated in Ref. [65]. One then gets $R = 4.6$ nm, a realistic value for the radius of curvature of a very good tip (Nanosensors guarantees $R < 7$ nm).

The force curve of Fig. 12c was obtained with these fitting parameters. The operating point corresponding to the images of Fig. 11 is indicated on the curve. The imaging force $F = -16$ pN is quite small, in agreement with the large tip–surface distance $d_0 = 1.6$ nm.

The fit of the $\Delta f(z)$ spectrum of Fig. 12b is also very good. The $P_1$ and $P_2$ parameters are close but different from those obtained on KBr, which is the

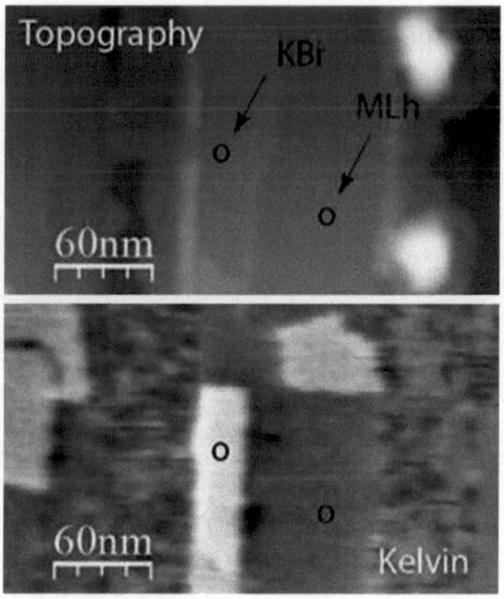


**Fig. 11** Image obtained on the sample of Fig. 2. $A_0 = 2$ nm, $\Delta f = -10$ Hz, $f_0 = 277.672$ Hz, $f_{mod} = 1$ kHz, $U_{mod} = 2$ V

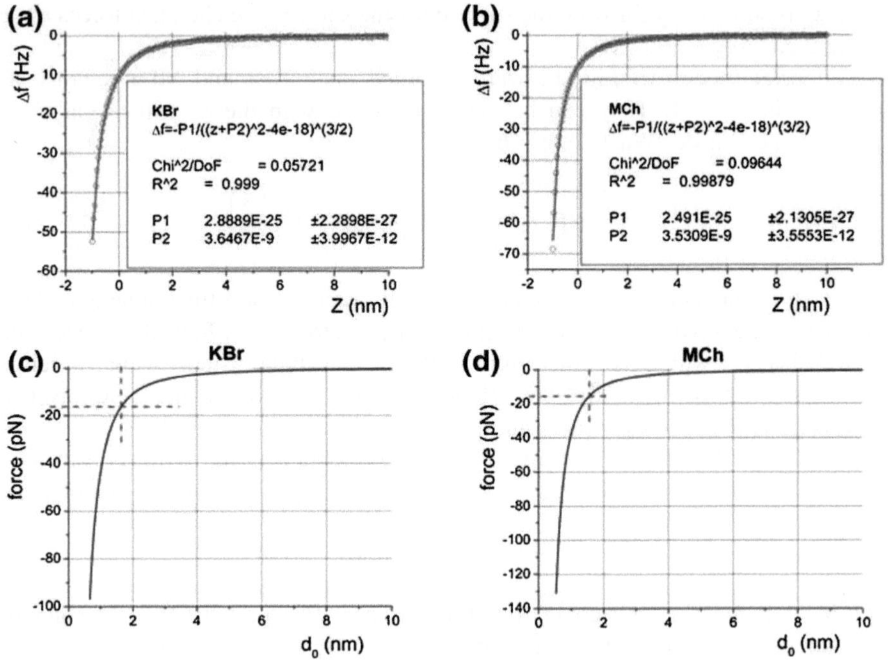


**Fig. 12** $\Delta f(z)$ spectra measured on the image of Fig. B.1 on **a** KBr and **b** MLh. *Black circles* experimental data, *red curves* fitting curves according to expression (1). Force curves calculated with the resulting fitting parameters on **c** KBr and **d** MLh (color figure online)

consequence of the presence of the molecular layer. Note that expression (5) is not fully adapted to describe this situation. The corresponding force curve is shown in Fig. 12d. The imaging force is nearly equal to the force used on KBr.

Note that there are two important sources of experimental uncertainty in these measurements: the values of $k$ and $A_0$. Nevertheless, they demonstrate that the imaging conditions adopted for the KPFM experiments described in this chapter correspond to an interaction regime where only van der Waals and electrostatic forces are involved.

## References

1. Pfeiffer, O., Gnecco, E., Zimmerli, L., Maier, S., Meyer, E., Nony, L., Bennewitz, R., Diederich, F., Fang, H., Bonifazi, D.: Force microscopy on insulators. Imaging of organic molecules. J. Phys. Conf. Ser. **19**, 166 (2005)
2. Kunstmann, T., Schlarb, A., Fendrich, M., Wagner, T., Möller, R., Hoffmann, R.: Dynamic force microscopy study of 3,4,9,10-perylenetetracarboxylic dianhydride on KBr(001). Phys. Rev. B **71**, 121403 (2005)

3. Burke, S.A., Ji, W., Mativetsky, J.M., Topple, J.M., Fostner, S., Gao, H.-J., Guo, H., Grütter, P.: Strain Induced dewetting of a molecular system: bimodal growth of PTCDA on NaCl. Phys. Rev. Lett. **100**, 186104 (2008)
4. Dienel, T., Loppacher, C., Stefan, C., Mannsfeld, B., Forker, R., Fritz, T.: Growth-mode-induced narrowing of optical spectra of an organic adlayer. Adv. Mater. **20**, 959 (2008)
5. Burke, S.A., LeDue, J.M., Miyahara, Y., Ji, W., Topple, J.M., Fostner, S., Grütter, P.: Determination of the local contact potential difference of PTCDA on NaCl. A comparison of techniques. Nanotechnology **20**, 264012 (2009)
6. Schütte, J., Bechstein, R., Rohlfing, M., Reichling, M., Kühnle, A.: Cooperative mechanism for anchoring highly polar molecules at an ionic surface. Phys. Rev. B **80**, 205421 (2009)
7. Glatzel, Th, Zimmerli, L., Koch, S., Kawai, S., Meyer, E.: Molecular assemblies grown between metallic contacts on insulating surfaces. Appl. Phys. Lett. **94**, 063303 (2009)
8. Pawlak, R., Nony, L., Bocquet, F., Oison, V., Sassi, M., Debierre, J.-M., Loppacher, C., Porte, L.: Supramolecular assemblies of 1,4-benzene diboronic acid on KCl(001). J. Phys. Chem. C **114**, 9290 (2010)
9. Such, B., Trevethan, T., Glatzel, T., Kawai, S., Zimmerli, L., Meyer, E., Shluger, A.L., Amijs, C.H.M., de Mendoza, P., Echavarren, A.M: Functionalized truxenes: adsorption and diffusion of single molecules on the KBr(001) surface. ACS nano **4**, 3429 (2010)
10. Trevethan, T., Such, B., Glatzel, T., Kawai, S., Shluger, A.L., Meyer, E., de Mendoza, P., Echavarren, A.M.: Organic molecules reconstruct nanostructures on ionic surfaces. Small **7**, 1264 (2011)
11. Hinaut, A., Lekhal, K., Aivazian, G., Bataillé, S., Gourdon, A., Martrou, D., Gauthier, S.: NC-AFM study of the adsorption of hexamethoxytriphenylene on KBr(001). J. Phys. Chem. C **115**, 13338 (2011)
12. Barth, C., Gingras, M., Foster, A.S., Gulans, A., Félix, G., Hynninen, T., Peresutti, R., Henry, C.R.: Two-dimensional nanostructured growth of nanoclusters and molecules on insulating surfaces. Adv. Mater. **24**, 3228 (2012)
13. Kittelmann, M., Rahe, P., Nimmrich, M., Hauke, C.M., Gourdon, A., Kühnle, A.: On-surface covalent linking of organic building blocks on a bulk insulator. ACS Nano **5**, 8420 (2011)
14. Kittelmann, M., Rahe, P., Gourdon, A., Kühnle, A.: Direct visualization of molecule deprotonation on an insulating surface. ACS Nano **6**, 7406 (2012)
15. Bocquet, F., Nony, L., Mannsfeld, S.C.B., Oison, V., Pawlak, R., Porte, L., Loppacher, Ch.: Inhomogeneous relaxation of a molecular layer on an insulator due to compressive stress. Phys. Rev. Lett. **108**, 206103 (2012)
16. Rahe, P., Kittelmann, M., Neff, J.F., Nimmrich, M., Reichling, M., Maass, P., Kühnle, A.: Tuning molecular self-assembly on bulk insulator surfaces by anchoring of the organic building blocks. Adv. Mater. **25**, 3948 (2013)
17. Schwarz, A., Gao, D.Z., Lämmle, K., Grenz, J., Watkins, M.B., Shluger, A.L., Wiesendanger, R.: Determining adsorption geometry, bonding, and translational pathways of a metal–organic complex on an oxide surface: co-salen on NiO(001). J. Phys. Chem. C **117**, 1105 (2013)
18. Kitamura, S., Iwatsuki, M.: Observation of the 7 × 7 reconstructed structure on the silicon (111) surface using ultra high vacuum noncontact atomic force microscopy. Appl. Phys. Lett. **72**, 3154 (1998)
19. Glatzel, T., Sadewasser, S., Lux-Steiner, M.C.: Amplitude or frequency modulation-detection in Kelvin probe force microscopy. Appl. Surf. Sci. **210**, 84 (2003)
20. Zerweck, U., Loppacher, C., Otto, T., Grafström, S., Eng, L.M.: Accuracy and resolution limits of Kelvin probe force microscopy. Phys. Rev. B **71**, 125424 (2005)
21. Barth, C., Foster, A.S., Henry, C.R., Shluger, A.: Recent trends in surface characterization and chemistry with high resolution scanning force methods. Adv. Mat. **23**, 477 (2011)
22. Sadewasser, S., Glatzel, T.: Kelvin Probe Force Microscopy: Measuring and Compensating Electrostatic Forces. Springer, Heidelberg (2011)
23. Zerweck, U., Loppacher, C., Otto, T., Grafström, S., Eng, L.M.: Kelvin probe force microscopy of $C_{60}$ on metal substrates : towards molecular resolution. Nanotechnology **18**, 084006 (2007)

24. Nikiforov, M.P., Zerweck, U., Milde, P., Loppacher, C., Park, T.-H., Uyeda, H.T., Therien, M. J., Eng, L., Bonnell, D.: The effect of molecular orientation on the potential of porphyrin-metal contacts. Nanoletters **8**, 110 (2008)
25. Zerweck, U., Loppacher, C., Eng, L.M.: Ordered growth and local workfunction measurements of tris(8-hydroxyquinoline) aluminium on ultrathin KBr films. Nanotechnology **17**, S107 (2006)
26. Léoni, T., Guillermet, O., Walch, H., Langlais, V., Scheuermann, A., Bonvoisin, J., Gauthier, S.: Controlling the charge state of a single redox molecular switch. Phys. Rev. Lett. **106**, 216103 (2011)
27. Barth, C., Henry, C.R.: Kelvin probe force microscopy on surfaces of UHV cleaved ionic crystals. Nanotechnology **17**, S155 (2006)
28. Barth, C., Henry, C.R.: Surface double layer on (001) surfaces of alkali halide crystals: a scanning force microscopy study. Phys. Rev. Lett. **98**, 136804 (2007)
29. Egberts, P., Filleter, T., Bennewitz, R.: A Kelvin probe force microscopy of charged indentation-induced dislocation structures in KBr. Nanotechnology **20**, 264005 (2009)
30. Hinaut, A., Pujol, A., Chaumeton, F., Martrou, D., Gourdon, A., Gauthier, S.: An NC-AFM and KPFM study of the adsorption of a triphenylene derivative on KBr(001). Beilstein J. Nanotechnol. **3**, 221 (2012)
31. Neff, J.L., Milde, P., Perez Leon, C., Kundrat, M.D., Eng, L.M., Jacob, C.R., Hoffmann-Vogel, R.: Epitaxial growth of pentacene on alkali halide surfaces studied by Kelvin probe force microscopy. ACS Nano **8**, 3294 (2014)
32. Palleau, E., Ressier, L., Borowik, Ł., Mélin, T.: Numerical simulations for a quantitative analysis of AFM electrostatic nanopatterning on PMMA by Kelvin force microscopy. Nanotechnology **21**, 225706 (2010)
33. Baier, R., Leendertz, C., Lux-Steiner, MCh., Sadewasser, S.: Toward quantitative Kelvin probe force microscopy of nanoscale distributions. Phys. Rev. B **85**, 165436 (2012)
34. Nony, L., Foster, A., Bocquet, F., Loppacher, C.: Understanding the Atomic-scale contrast in Kelvin probe force microscopy. Phys. Rev. Lett. **103**, 53 (2009)
35. Sadeghi, A., Baratoff, A., Ghasemi, S.A., Goedecker, S., Glatzel, T., Kawai, S., Meyer, E.: Multiscale approach for simulations of Kelvin probe force microscopy with atomic resolution. Phys. Rev. B **86**, 075407 (2012)
36. Giessibl, F.J.: Forces and frequency shifts in atomic-resolution dynamic-force microscopy. Phys. Rev. B **56**, 16010 (1997)
37. Sadewasser, S., Lux-Steiner, MCh.: Correct height measurement in noncontact atomic force microscopy. Phys. Rev. Lett. **91**, 266101 (2003)
38. Fuchs, F., Grévin, B., Bocquet, F., Nony, L., Loppacher, C.: Correct height measurements by Kelvin probe force microscopy: poly(3-dodecylthiophene) on highly oriented pyrolytic graphite. Phys. Rev. B **88**, 205423 (2013)
39. Rahe, P., Lindner, R., Kittelmann, M., Nimmrich, M., Kühnle, A.: From dewetting to wetting molecular layers: $C_{60}$ on $CaCO_3$(10-14) as a case study. Phys. Chem. Chem. Phys. **14**, 6544 (2012)
40. Barth, C., Hynninen, T., Bieletzki, M., Henry, C.R., Foster, A.S., Esch, F., Heiz, U.: AFM tip characterization by Kelvin probe force microscopy. New J. Phys. **12**, 093024 (2010)
41. Materials Studio v5.5.0.0, Accelrys Software Inc.: San Diego, CA
42. Documentation of Materials Studio v5.5.0.0, Accelrys Software Inc., San Diego, CA. http://accelrys.com/products/datasheets/compass.pdf
43. Sun, H.: COMPASS: an ab initio force-field optimized for condensed-phase applications: overview with details on alkane and benzene compounds. J. Phys. Chem. B **102**, 7338 (1998)
44. Zhao, L., Liu, L., Sun, H.: Semi-ionic model for metal oxides and their interfaces with organic molecules. J. Phys. Chem. C **111**, 10610 (2007)
45. Trevethan, T., Shluger, A.L.: Building blocks for molecular devices: organic molecules on the MgO (001) surface. J. Phys. Chem. C **111**, 15375 (2007)
46. Gomez-Monivas, S., Froufe, L.S., Caamano, A.J., Saenz, J.J.: Electrostatic forces between sharp tips and metallic and dielectric samples. Appl. Phys. Lett. **79**, 4048 (2001)

47. Sacha, G.M., Saenz, J.J.: Cantilever effects on electrostatic force gradient microscopy. Appl. Phys. Lett. **85**, 2610 (2004)
48. Elias, G., Glatzel, T., Meyer, E., Schwarzman, A., Boag, A., Rosenwaks, Y.: The role of the cantilever in Kelvin probe force microscopy measurements. Beilstein J. Nanotechnol. **2**, 252 (2011)
49. Ma, Z.M., Kou, L., Naitoh, Y., Li, Y.J., Sugawara, Y.: The stray capacitance effect in Kelvin probe force microscopy using FM, AM and heterodyne AM modes. Nanotechnology **24**, 225701 (2013)
50. Sadeghi, A., Baratoff, A., Goedecker, S.: Electrostatic interactions with dielectric samples in scanning probe microscopies. Phys. Rev. B **88**, 035436 (2013)
51. Kelvin, L., Fitzgerald, G., Francis, W.: Contact electricity of metals. Phil. Mag. J. Sci. **46**, 82 (1898)
52. Zisman, W.A.: A new method of measuring contact potential differences in metals. Rev. Sci. Instr. **3**, 367 (1932)
53. Olsson, L., Lin, N., Yakimov, V., Erlandsson, R.: A method for in situ characterization of tip shape in ac-mode force microscopy using electrostatic interaction. J. Appl. Phys. **84**, 4060 (1998)
54. Kantorovich, L.N., Foster, A.S., Shluger, A.L., Stoneham, A.M.: Role of image forces in non-contact scanning force microscope images of ionic surfaces. Surf. Sci. **445**, 283 (2000)
55. Kantorovich, L.N., Livshits, A.I., Stoneham, M.: Electrostatic energy calculations for the interpretation of scanning force microscopy experiments. J. Phys. Condens. Matter **12**, 795 (2000)
56. Hynninen, T., Foster, A.S., Barth, C.: Polarized tips or surfaces: consequences in Kelvin probe force microscopy. e-J. Surf. Sci. Nanotech. **9**, 6 (2011)
57. Borowik, L., Kusiaku, K., Théron, D., Mélin, T.: Calculating Kelvin force microscopy signals from static force fields. Appl. Phys. Lett. **96**, 103119 (2010)
58. Hudlet, S., Saint Jean, M., Guthmann, C., Berger, J.: Evaluation of the capacitive force between an atomic force microscopy tip and a metallic surface. Eur. Phys. J. B **2**, 5 (1998)
59. Durand, E.: Electrostatique et magnétostatique. Masson, Paris (1953)
60. Smythe, W.R.: Static and Dynamic Electricity, 3rd edn, p. 227. Mc Graw Hill, New York (1968)
61. Jeans, J.H.: The Mathematical Theory of Electricity and Magnetism, 5th edn, p. 291. Cambridge University Press, Cambridge (1927)
62. Terris, B.D., Stern, J.E., Rugar, D., Mamin, H.J.: Contact electrification using force microscopy. Phys. Rev. Lett. **63**, 2669 (1989)
63. Gauthier, S., Martrou, D., in preparation
64. Aimé, J.P., Boisgard, R., Nony, L., Couturier, G.: Nonlinear dynamic behavior of an oscillating tip-microlever system and contrast at the atomic scale. Phys. Rev. Lett. **82**, 3388 (1999)
65. Bergström, L.: Hamaker constants of inorganic materials. Adv. Coll. Int. Sci. **70**, 125 (1997)

# Index





P. Moriarty and S. Gauthier (eds.), *Imaging and Manipulation of Adsorbates Using Dynamic Force Microscopy*, Advances in Atom and Single Molecule Machines,
DOI 10.1007/978-3-319-17401-3

MIX
Papier aus verantwortungsvollen Quellen
Paper from responsible sources
FSC® C105338

If you have any concerns about our products,
you can contact us on
ProductSafety@springernature.com

In case Publisher is established outside the EU,
the EU authorized representative is:
Springer Nature Customer Service Center GmbH
Europaplatz 3, 69115 Heidelberg, Germany

Printed by Libri Plureos GmbH
in Hamburg, Germany

MIX
Papier aus verantwortungsvollen Quellen
Paper from responsible sources
FSC® C105338

If you have any concerns about our products,
you can contact us on
**ProductSafety@springernature.com**

In case Publisher is established outside the EU,
the EU authorized representative is:
**Springer Nature Customer Service Center GmbH**
**Europaplatz 3, 69115 Heidelberg, Germany**

Printed by Libri Plureos GmbH
in Hamburg, Germany